Acknowledgements

The APS has drawn on the knowledge and experience of many members who have contributed to APS Guidance documents in the past and to the consultations that have continued throughout the preparation of this Guide.

This Guide to Design Risk Management has drawn upon experience of these processes gained over the last eleven years of practice under the CDM Regulations, together with the considerable analysis and comment of the APS team monitoring the development of the new set of regulations and approved Code of Practice over the past two years.

Recognition of this considerable input from membership experience is acknowledged by the APS and the Editor.

Particular thanks are due to those who formed the editorial team: Greg Brown, Pauline Jones, Graham Leech, Brian Law, Fraser Cook, Nick Charlton Smith (Editor) and to all those who have commented on draft text.

The following organisations also contributed to the development of this guide: Human Technology Systems Ltd, Faber Maunsell Ltd, CDM (Scotland) Ltd, The Charlton Smith Partnership.

Foreword

This Guide has been prepared to help Designers who are concerned to get things right this time around when dealing with the CDM Regulations. The process of Design Risk Management, insofar as it concerns Health and Safety in construction, is an integral part of the design process and can make a substantial contribution to achieving Client and design team targets on projects if seen in this light. It can also make a big difference to the lives, health and well-being of those who work on the buildings and other structures with which we are involved.

The Construction (Design and Management) Regulations 2007 present a major challenge and opportunity to our industry to try to drive down the human costs of construction while, at the same time, helping to achieve completions that are on time, to budgeted costs and to the required quality standards. Designers have a major part to play in meeting these challenges and taking the opportunities that are presented, for it is their work that determines what has to be done by those who construct, clean, maintain, repair and demolish or dismantle our constructions. Without their full engagement in this process of reducing risk and achieving safer construction and maintenance, little will be achieved; with their engagement, a lot can be achieved and we look forward to seeing the results of this over the next decade as we all get to grips with creating a safer construction industry in the United Kingdom.

At this stage we can only surmise how the new regulations will affect practice – but this Guide has tried to draw out the implications of the new regulations and Approved Code of Practice to see what they really mean for Designers who have to apply them. We hope that you will find it useful and that you will also take the opportunity to let the APS know about your practice experiences of working under the new CDM regime – after all, we need feedback to be able to develop good practice standards. We hope that this Guide is a starting point in that process.

Association for Project Safety: Register of Designers

The APS's Register of Designers is open to all designers who can demonstrate a satisfactory level of experience, knowledge and understanding of construction health and safety risk management in design. Members of this Register, which is open to all who can satisfy the entrance requirements and accept the Association's CPD requirements, will receive a wide range of services from APS including further guidance through Practice Notes, Newsletters, Legal Advice Service and Practice Help-line as well as CPD provision that focuses on construction industry and CDM-related health and safety issues.

The Register is intended to be an indicator of capability in Design Risk Management but also a means of distilling and developing good practice and enabling those involved in it to engage with other industry players who are concerned about construction health and safety. In this way, those who work in a wide range of roles that have to engage with health and safety can learn from each other and widen their knowledge and experience of the practical issues that can help construction become a safer industry.

More information on the Register of Designers can be obtained from APS.

Association for Project Safety
Stanhope House,
12, Stanhope Place,
Edinburgh,
EH12 5HH

Tel: 08456 121 290

Website: http://www.aps.org.uk/

Contents

CONTENTS

Starting Out

1.1 What This Guide Does/Does Not Do

This Guide is intended to help Designers steer an effective path through the duties and responsibilities that they have to discharge under the Construction (Design and Management) Regulations 2007 (CDM Regulations). By Designers we mean, of course, all those who make any design decisions on projects, including 'specifiers', 'calculators', sometimes Clients, as well as Designers in the more 'traditional' sense.

The Guide is founded on the proven fact that elimination, reduction and effective management of Health and Safety risks on construction projects can have positive benefits in terms of project cost, delivery and quality targets (see, for example, data on CE Demonstration Projects). In addition, the (increased) co-operation and co-ordination that CDM requires can also bring spin-off benefits to all those involved in the project, so that time spent is used more efficiently and redundant, repeat or remedial work is reduced or avoided altogether.

The Guide encapsulates what the Association for Project Safety (APS) considers, at this point in time, to be good practice in Health and Safety-related Design Risk Management (DRM), and it addresses the implications of the responsibilities and duties of Designers under the CDM Regulations. However, underpinning the Guide is the assumption that these Health and Safety-related issues can, and should, be integrated into the holistic design process and so contribute to the effective delivery of projects.

The Guide covers:

- The need for Clients to be aware of their duties
- The capability, competence and resources issues that Designers need to consider
- The capability of design sub-consultants and other organisations to carry out their design work effectively
- The role of co-operation and co-ordination in DRM
- The use of information to inform design and planning
- The process of DRM
- Issues relating to residual Health and Safety-related information

Individual Designers and firms of Designers using this Guide are assumed to be basically competent in Health and Safety DRM, to have read the Regulations and Approved Code of Practice (ACoP), and to have the necessary capability to work effectively on the particular projects with which they are concerned (aspects of which are covered in Section 3).

The Guide does not generally repeat material in the CDM Regulations and the associated ACoP, but tries to provide practical guidance to help Designers meet the requirements of both. It is assumed that Designers will have to hand and will have read both of these documents.

1.2   Notifiable and Non-Notifiable Projects

Under CDM there are two types of construction projects: those that are notifiable to the Health and Safety Executive (HSE) and those that are not. The cut-off between the two is either 30 days or more of construction work or 500 (normal) working days or more of construction work. Projects above either of these limits are notifiable. (It is worth noting that 'working days' in this context refers to a normal shift of work carried out by one person.)

On Notifiable Projects a CDM Co-ordinator (CDM-C) is available to provide Clients with guidance and assistance, but on Non-Notifiable Projects someone else may need to do this. For this reason, the APS considers that Non-Notifiable Projects could be more onerous for Designers in a number of respects because Clients are likely to look to them for advice and assistance with the discharge of their duties.

Notifiable Projects: Designers' duties are clearly defined within the CDM Regulations and include making sure that the Client is aware of his or her duties – probably the most important of which is the appointment of a CDM-C who is then available to the Client to provide explanations of duties, help establish management arrangements, implement and facilitate Health and Safety-related project co-ordination, and to generally assist with the discharge of Client duties.

Non-Notifiable Projects: Designers' duties are the same as for the Notifiable Project but the Client has no-one appointed to advise and assist with explanations and discharge of duties and no-one to help with project co-ordination – all of which the Client may well need.

Good practice suggests that the 'prime' or 'lead' Designer may be the most appropriate person to provide this assistance and, at the very least, to ensure that the project is effectively co-ordinated in terms of Health and Safety risk management. This assistance could cover:

- Explanations of Client duties and their implications, including in relation to the start of construction work
- Demonstrating competence and capability to the Client and enabling the Client to understand such issues
- Explaining the need for co-ordination and assisting with or carrying out the necessary project co-ordination on behalf of the Client
- Helping the Client to identify the information needed for the project and assisting with its provision (what to provide, what to seek) and dissemination, where appropriate
- Gathering and identifying existing (residual) risk information to assist Designers and Contractors with design and planning for construction work
- Compiling, circulating and keeping information on existing and residual risks, especially residual information from the designs prepared by other Designers
- Providing the Client with relevant information at the end of a project so that any existing Health and Safety File (the File) or Asbestos Register can be updated [see Regulation 17(3)(b)]

On the other hand, where lead Designers are concerned about their capability to provide this assistance, Clients may be advised to appoint a professional to do this (perhaps someone who provides CDM-C services on Notifiable Projects). This may be especially important on projects where the levels of risk and project complexity are high and extend beyond that Designer's experience and competence. (Projects may be short in duration but complex and involve significant or difficult-to-manage risks – Non-Notifiable does not, inter alia, mean low risk!) In either case, there could be a cost implication for the services required and Designers may need to review implications in terms of time and fees for the additional services if they choose to provide them.

The Guide is written so that Designers can understand the implications for them of both Non-Notifiable and Notifiable Projects, with the differences being made clear at each stage.

1.3 Design Risk Management (DRM)

It is normal as part of any design process for any inherent risks to be identified and strategies to eliminate or manage those risks integrated into the developing design. In this Guide, the focus is on the management of Health and Safety issues that can affect those working on the project or those who can be affected by the work itself. So, DRM is seen as the Health and Safety risk management process that competent Designers will engage with as an integral part of the overall design process.

The process of DRM involves identifying hazards or hazardous activities and any associated risks relating to the intended construction work (building structures, maintaining, cleaning, using – in certain circumstances – and removing them). These can be associated with the sites(s), the existing structures on which work will be carried out and their surroundings, as well as the hazards and risks that are created by the designs that are produced for the new structures or constructions. Designers are then required to eliminate or minimise the identified risks by changing or modifying aspects of designs, and to provide information about any resultant *significant* residual risks so that these can be managed safely during the construction process and during maintenance, cleaning and eventual demolition.

Underpinning this DRM process is the recognition that making structures safer for those who work on them, or are affected by construction of them, will have beneficial spin-offs in terms of project delivery, cost and quality targets – a fact proved many times over in recent years.

The DRM process includes providing information for:

- Other Designers and for Contractors so that they can design, plan and prepare effectively
- Tendering or pricing purposes (so that risk management is properly taken into account in construction)

- Ongoing construction Health and Safety through the use of the Health and Safety File

It is important to note that the DRM process assumes that the Contractors appointed or the Designer engaged to carry out work on a project will be competent, and therefore that the residual design risk information that Designers need to produce will relate only to those risks that are not easily apparent, are seen to be unusual or difficult to manage, and of which a competent Contractor or Designer would not otherwise be aware.

DRM is an integral part of both singular and iterative design processes. On Notifiable Projects, DRM will be co-ordinated by a lead Designer or by the CDM-C who will work with Designers to:

- Try to minimise risk in construction to levels compatible with other project aims and objectives
- Ensure that information flows between all parties involved in the design and construction so that Health and Safety risk management is effective

On Non-Notifiable Projects where there is no CDM-C, someone else will have to ensure that this is done so that the Client's duties are fully discharged and the team does all that is practicable to ensure the Health and Safety of those working on the project or those potentially affected by it.

Reminder:

Designer CDM duties apply from Design Day 1 on every project

1.4 Starting Out

The starting point for any Designer, in discussion with the Client, must be to assess whether or not the construction phase is likely to involve more than 30 days, or 500 person days, of construction work.

Where there is no reasonable certainty that the project will be less than 30 days or 500 (normal) working days of construction work, the APS recommends that Designers advise Clients that the project should be Notified and a CDM-C appointed. This is strongly recommended when the work that has to be carried out is likely to include inherently risky or complex construction operations such as significant demolition or underpinning, perhaps alongside other activities.

When a Project is Notifiable, the Designer will need to:

- Check that the Client is aware of his duties
- Ensure that the Client understands the essential duty to appoint a CDM-C before design work commences (*see Section 2*)
- Make clear that, until the Client makes the CDM-C appointment (in writing), the HSE will regard him (or her) as discharging the CDM-C's duties – with all the consequent liabilities that these involve

Once appointed, the CDM-C will be able to advise and assist with all other Client duties including further discussions about their implications.

> *Before doing anything else check that the Client is aware of his CDM duties*

When a Project is Non-Notifiable, the Client may expect the Designer not only to advise on the Client duties that have to be discharged, but also to assist with aspects of discharging them.

Clients who have knowledge and experience of construction work may need no assistance at all with Non-Notifiable Projects – but many, who have no experience or only limited experience, will expect their professional team to help keep them on track. With complex projects that may have up to 16 people working on them for up to 30 days, they may need guidance on and help with discharging their duties. If a Designer provides this assistance, it will increase Designer time and costs that may need to be reflected in the fees and other charges. However, these can only be realistic if Designers understand what they might need to do and the time that these activities will take.

On the other hand, the Designer might advise the appointment of someone else (who, perhaps, provides CDM-C services on Notifiable Projects) to provide this service even though such an appointment is not required under the CDM Regulations. Either way, there may be an additional cost to the Client for this service – but this is offset by the comfort that the Client will derive in relation to the effective discharge of duties.

This Guide works from the premise that this advice and assistance is likely to be needed, and provides guidance on the effective discharge of these additional services. A Client Guidance leaflet titled 'CDM 2007: What Clients Need To Do' is available from the APS to assist Designers with advising Clients of their duties.

Client Awareness of Duties

2.1 Overview

This section of the Guide deals with the Client's duties and how Designers can check that Clients are aware of them.

The CDM Regulations impose a legal obligation not to start design work on a construction project (other than 'initial design work' – discussed below) unless a Client is aware of the duties laid on them by those regulations and, on Notifiable Projects, has appointed a CDM-C who can, of course, explain duties to a Client. However, on Non-Notifiable Projects, while there is no requirement in the regulations for the Designer to do so, the only way to be sure that Clients are aware of their duties may be for Designers themselves to provide a full explanation. This is the view taken in this Guide, and this section is intended to help Designers provide that explanation.

2.2 Notifiable or Non-Notifiable?

The first thing that will need to be discussed and decided with the Client is whether or not the project is Notifiable. This issue has been covered in Section 1 but it is worth reminding ourselves that, where it is not clear which category the project will fall into, the APS recommends that Clients are advised to treat it as Notifiable, especially if the project contains high-risk activities and/or is complex.

2.3 Clients: 'Aware' or 'Not Aware' of Duties?

Start of design work: As a Designer must not start design work (other than initial design work) until satisfied that a Client is aware of his/her duties (and, on a Notifiable Project, not until a CDM-C has been appointed), the second issue for Designers to check is that Clients are aware of their duties and their implications. Checks can be made by:

- Considering previous experience of working with that Client
- Discussing the CDM Regulations and establishing whether or not the Client is aware of the revised duties
- Asking Clients to describe their understanding of the duties imposed on them by the new regulations

Initial Design Work: Preliminary design work, planning or other preparation for construction work

The regulations and ACoP are reasonably clear on this issue:

Reg 14: '... *appoint ... as soon as is practicable after ... initial design work, or other preparation for construction work has begun*'.

Reg 18: '... *no Designer shall commence work (other than initial design work) ... unless a CDM co-ordinator has been appointed ...*'

ACoP paras 66 and 86 in relation to appointment of the CDM-C state: '*Early appointment is crucial for effective planning and establishing management arrangements from the start ...* **The co-ordinator needs to be in a position to be able to co-ordinate design work and advise on the suitability and compatibility of designs, and therefore should be appointed before significant detailed design work begins. Significant detailed design work includes preparation of the initial concept design and implementation of any strategic brief. As a scheme moves into the detailed design stage, it becomes more difficult to make fundamental changes that eliminate hazards and reduce risks associated with early design decisions**'.

The APS view of the implications of these statements, taken in the context of the Regulations, the ACoP as a whole, and together with the fundamental reasons for early appointment of the CDM-C is that they indicate that initial design work *should not be inferred* to include initial designs for the project, competition designs or more detailed designs but should relate to the very earliest explorations that indicate that a project has the potential to match an initial Client brief and notions of cost. Appointment at any later stage would frustrate the purpose of the appointment of the CDM-C, information flow, co-ordination and competence requirements and place the Client in the position of CDM-C for a critical phase of DRM.

If there is any doubt, a few appropriate questions should help to identify whether Clients really understand what is required of them – especially with regard to the management arrangements, information provision, start of works on site and welfare provisions.

Who is checked? The checks are relatively straightforward if the Client is a small organisation or an individual but, where large organisations are involved (for example, Highways Agency, Local Authority, Government Ministry, Statutory Authority or major developer), checks on awareness of duties will need to be applied to the particular individuals representing the Client – not simply the 'organisation as a whole'.

Explanation of duties by CDM-C: On a Notifiable Project, the Designer must stress the need for the early appointment of a CDM-C and explain that the CDM-C will assist with explanations and clarification of all other Client duties.

If a CDM-C has been appointed and the HSE notified, the Designer can assume that the Client is aware of his/her duties under the regulations and that any required explanations have been carried out by the CDM-C.

It is worth reminding the Client, however, that any delay in the appointment of the CDM-C will place the duties of the CDM-C on the Client, and the Client will also be deprived of the 'empowered Health and Safety advisor who is pivotal in ensuring an effective project team' that the regulations require.

Explanation of duties by Designer: On a Non-Notifiable Project, all the necessary explanations will fall to the Designer to provide unless the Client is already aware of the duties that the regulations impose.

If it is evident that a Client does not understand what they are required to do, the APS recommends that Designers should explain these duties carefully, discuss the implications of the regulations, and provide information on good practice in discharging them. This can be done in a number of ways, but it may be helpful to give them a copy of the APS leaflet '*CDM 2007: What Clients Need to Do*' and then discuss the contents with them using the leaflet as an agenda for explanation and discussion.

Once it has been confirmed that a Client is aware of duties under the regulations (and the implications of these) and that a CDM-C has been appointed, the Designer can start design work.

2.4 Advantages of CDM Risk Management for the Client

To be able to explain the duties and their implications to Clients, Designers have to be clear on what they are and what needs to be done. The next part of this Section therefore concentrates on expanding the brief details contained in the Client leaflet.

When talking to a Client about the requirements of the CDM Regulations, it may be helpful to stress the advantages of improved information flows, co-ordination and improved project planning, together with reductions in Health and Safety risks, in terms of fundamental Client objectives to achieve hand-over on time, at the budgeted cost, and to the required quality standards.

If Clients are convinced that the regulations (and what they, as Clients, have to do) are in the best interests of their project, they are more likely to be persuaded to carry out their duties effectively. This will have advantages for all involved. These advantages include:

- Improved planning and management
- Reduced costs, delays and bad publicity from accidents or ill health

Improved planning and management: No-one needs the stress of a project that is not running smoothly. Good planning and management are key factors in avoiding this situation, helping to achieve completion on time and within budget. Good quality workmanship is more readily achieved when projects are well planned and managed. In addition, information from today's design and planning can reduce the likelihood of accidents occurring during future works.

Reduced costs, delays and bad publicity from accidents or ill health: Every Health and Safety incident on site has a consequence and a potential cost. Every accident is an impediment to the smooth completion of the project and achievement of its objectives. Each one imperils the achievements of the whole team and the Client's targets so, wherever and whenever possible, these should be eliminated by improved project design and planning. Health and Safety Risk Management, which is what the CDM regulations are all about, is a 'sub-set' of overall Construction Risk Management and therefore should be a key concern of all involved – Client, Designers, CDM-C and Contractors. The regulations are therefore a means to an end, not the end itself.

Clients need to be made fully aware that:

- Accidents and ill health on site can cost a lot of time and money, reduce morale, and will often affect standards of workmanship
- Poor Health and Safety performance can attract HSE interest
- Consequent 'Improvement' or 'Prohibition' Notices, whether or not there is an accident, can have significant time and cost implications
- The local and national press will take great pleasure in publishing details of accidents and, in this image-conscious age, it is just not true that 'any publicity is good publicity'!

Reduced whole life costs of building or structure: Planning for safe construction, maintenance, cleaning, use as a workplace, and demolition can result in:

- Use of materials and constructions that need less maintenance/cleaning
- Provision of safe access for maintenance/cleaning with potential for reduced long-term costs
- Improved records and information on structures and constructions which will reduce the need for investigations for future alterations or extensions
- Design that facilitates eventual demolition and disposal

All can impact on whole life costs – bearing in mind that these costs will vastly outstrip construction costs. The modest additional costs of the effective application of CDM now can be recovered many times over in the life of a building or structure.

If construction costs are a factor of 1 in the overall costs of a building, then maintenance and building operational costs are 5 and life-time business costs are 200.

Time spent in design reducing costs in use can bring significant life-time cost benefits

2.5 Client Duties on All Projects

The following Client duties must be discharged on all projects, irrespective of size or complexity or the duration of the construction work:

Appointing the right people at the right time: The Client must appoint or engage people and organisations that are competent and capable of carrying out the work they have to do. This includes Designers, CDM-C and Contractors (see later section on competence). These people or organisations must be:

- Adequately resourced to carry out the specific project. This includes having sufficient appropriately skilled and trained personnel available and correct equipment where necessary
- Appointed or engaged early enough to carry out their duties. A key issue in carrying out any work effectively is having enough time to do it

It is not cost effective to delay appointments or engagements as the amount of work to be carried out will remain the same, whatever the time scale, but delays in appointments can often result in work having to be halted while the new appointee catches up on critical Health and Safety risk management issues.

> *On Notifiable Projects the CDM Co-ordinator will assist the*
> *Client with checks on competence and resources*

Management arrangements: The Client must make sure that management arrangements are in place throughout the project so that construction works can be carried out safely and without risk to health. These will include:

- Informing the team of any instructions from the Client relating to how the project is to be run, to the protection of others such as neighbours, the Clients' own workforce, and interaction with other projects
- Ensuring all members of the team understand clearly their roles, functions and responsibilities
- Setting up systems for good communication, co-ordination and co-operation between all members of the team to help achieve a satisfactory project completion
- Arranging for Contractors to confirm that Health and Safety standards on site, including welfare facilities, will be managed and monitored and how this will be carried out

Many of these arrangements will be made by members of the team, so the Client will need to ask them to explain what arrangements they have made to ensure they comply with the requirements. They should also be asked to explain how they will ensure that these arrangements are maintained throughout the project. It could be helpful to all concerned if the above arrangements were contained within a single document for easy reference.

The Client should periodically check that the arrangements are being implemented and updated as necessary.

On Notifiable Projects the CDM Co-ordinator will assist the Client with all of these arrangements – including advising on what they should consist of, if necessary putting them or helping to put them in place, and checking that they are being maintained

Allowing sufficient time: The Client must allow sufficient time for each stage of the project, including planning and mobilisation. The Client may need advice on appropriate time scales from Designers, Quantity Surveyors, project managers or Contractors.

On Notifiable Projects the CDM Co-ordinator will be able to advise the Client on time scales and mobilisation – in conjunction with other members of the project team

Providing information: The Client must provide any relevant information that is likely to be needed by those involved in the project to enable them to comply with their duties. This could include:

For Designers:

- A previous Health and Safety File, if available
- Asbestos Register and management plan
- Survey information, including adjacent land and premises

- Previous uses of the site
- Ground conditions, including contamination
- Information about the Client's on-site activities
- Client arrangements, deadlines or requirements that may impact on Health and Safety
- Any known hazards
- Proposed use, if intended as a workplace
- Any other relevant information identified by Designers that can be reasonably obtained by the Client

For Contractors:

- All of the above information (or appropriate sections of it)
- The minimum length of time they will be given for planning and preparing to start on site (mobilisation period)
- Any residual hazards identified during the design stage
- Any minimum Health and Safety standards set by the Client

Once again it would be helpful to all concerned if the above information was contained within a single document for easy reference. (*See Section 6 for further details of the Project Information Resource.*)

*On Notifiable Projects the CDM Co-ordinator will have a major role to play in ensuring that the **right information is provided to the right people at the right times** as part of the co-ordination process in DRM. The sooner the CDM Co-ordinator is appointed, the easier it will be for the Client to be assured that this duty is being effectively discharged*

Co-operation: The Client must co-operate with and seek the co-operation of all others involved in the project to make it easier for them to carry out their duties. This will include:

- Appointing people at the appropriate time to allow co-operation and co-ordination
- Not imposing restrictions that compromise others' compliance with their duties

- Making timely decisions
- Not unreasonably withholding information

Of course the duty to seek and provide co-operation applies to all those involved in a project.

Co-ordination of Client's work: The Client should co-ordinate their own work with all those involved to ensure the safety of those carrying out construction work and anyone affected by it. This is vital where the Client's operations are to continue on premises or on site while a construction project is taking place.

On Notifiable Projects the CDM Co-ordinator will assist the Client with this duty and help ensure that this is integrated into the whole project management process

Designing workplaces to comply with Health and Safety legislation: The Client should ask Designers to confirm that this is being carried out and must not impose any restrictions on Designers that would compromise their ability to do so.

On Notifiable Projects CDM Co-ordinators have to check that designs are suitable and compatible and cover workplace Health and Safety issues

Time to mobilise for construction work to start: The Client must give notice to the Principal Contractor (and any other direct Contractor appointees) of the time that will be allowed for planning and preparation between appointment and starting construction work. This is to ensure that sufficient time is available for them to organise the management of Health and Safety on the site and to enable welfare facilities to be provided prior to the start of construction work.

On Notifiable Projects the CDM Co-ordinator will help establish a suitable mobilisation period in conjunction with other members of the project team

Welfare facilities: A Client must not allow works to start on site unless satisfied that suitable welfare facilities will be provided by Contractors from Day 1 and must ensure that they are maintained throughout the construction phase. It may be helpful to suggest that the Client issues a formal authorisation to commence work once satisfied that suitable welfare facilities will be provided throughout the Construction Phase – the APS produces a Notice that can be used by Clients to do this.

On Notifiable Projects the CDM Co-ordinator will be able to establish that suitable welfare facilities will be in place and advise the Client that, once any other of their requirements have been met, construction work can start

Information for future projects: Although there is no legal obligation under the CDM Regulations for a Client to pass on or keep residual risk information from Non-Notifiable Projects, it is a sensible action to take where a building/structure already has a File or an Asbestos Register. Such information will enable a Client to comply with the duty to keep a Health and Safety File up to date and also to provide the basis for providing information for possible future projects.

2.6 Additional Duties for Notifiable Projects

When a project is Notifiable, the Client has all the duties listed above, together with further duties that relate to appointments and formal documentation.

Formal appointments: The CDM Regulations require the Client to appoint a CDM-C and a Principal Contractor and to make these appointments in writing.

Alert!!

Clients should be warned that, until they make these appointments, they will be legally liable for the associated duties themselves and will be deemed to be carrying them out. They should also be made aware that, if they don't carry them out adequately, they could face criminal prosecution by the HSE – especially if there is an accident

CDM Co-ordinator (CDM-C) to help Clients carry out their duties: The Client must appoint a CDM Co-ordinator (CDM-C) – who is their empowered, key project Health and Safety adviser – as soon as possible to advise and assist them and also to ensure that arrangements are made and implemented for the co-ordination of Health and Safety during the planning and preparation phase. The CDM-C may be a person or an organisation, depending on the size and complexity of the project.

The CDM-C must be appointed as soon as practicable after initial appraisal of project needs and objectives and before preparation of the initial concept design. This means appointing before significant initial design work begins (*see 2.3 above*). Clients *and* Designers must be clear that significant design work cannot be embarked upon until the CDM-C has been appointed.

Designers should emphasise to their Clients that the CDM-C can help with all the Client's duties, but this can only be effective if the appointment is made at the earliest stage. Until the CDM-C is appointed, the regulations assume that the Client will be discharging all of the CDM-C's duties.

Other duties: Clients have other duties on Notifiable Projects that are briefly mentioned below in case a Designer needs to explain to a Client what these are – but all would be dealt with, or the Client advised on such matters, by the CDM-C once appointed to the project.

Appointment of a Principal Contractor: Clients are required to appoint a Principal Contractor as soon as is practicable after the Client knows enough about a project to be able to select someone suitable for that type of construction work. Where it is not yet feasible to appoint the Principal Contractor that will undertake the actual construction phase of the project, there is nevertheless the opportunity to appoint someone who can advise the project team on construction and buildability issues as well as on welfare facilities, programme mobilisation time and the like. Early

As this appointment is only required on Notifiable Projects, the CDM Co-ordinator will be available to advise the Client on the early appointment of a Principal Contractor as well as on necessary changes to such appointments as the project develops

availability of this type of input, whether by the project PC or someone else at an earlier stage could benefit design, construction and costs in use of the structure.

Provision of information: The Client also has to provide the CDM-C with the pre-construction information that the project team will need (*see 2.5 above*) as well as information that may be needed for the Health and Safety File for the project so that the CDM-C can pass that information to the right people at the right times. The CDM-C will assist the Client in this, in conjunction with other members of the project team.

Start of construction phase: The Client has to ensure not only that adequate welfare facilities will be in place before construction work commences, but also that they will remain in place throughout the construction phase. The Client also has to ensure that a 'Construction Phase (Health and Safety) Plan' has been prepared by the appointed Principal Contractor (for the construction phase) before construction work can start. This plan has to be sufficient to ensure that the construction phase is planned, managed and monitored in such a way that construction work can be carried out safely.

On such a project the CDM-C will be available to advise and assist the Client with the discharge of these duties.

2.7   Confirmation of Understanding

Once it is clear that a Client understands their duties, this should be confirmed in writing – probably by letter from the Designer to the Client. However, if 'awareness' is assumed on the basis that a CDM-C has been appointed (see above), it would be sensible to obtain written confirmation that this appointment has, in fact, been made by asking for a copy of the F10 or equivalent notice to the HSE. This will include the name of the CDM-C, as well as a declaration signed by the Client or his representative (from within his organisation) that he is aware of his duties under the CDM Regulations. This is another small, but important, change to the CDM regime – to ensure that Clients do know what is required of them on Notifiable Projects.

2.8   Overall: When Talking to Clients

Do...

- Determine straight away whether the project is Notifiable or Not-Notifiable and advise the Client accordingly
- Make sure that Clients are aware of their duties before commencing design work
- Present duties in a positive and project-beneficial light when explaining them to Clients
- Stress the value of good management and the benefits of the CDM Regulations in terms of achieving 'on time, on cost and on quality' delivery of projects and not just the Health and Safety of the construction workforce
- Advise Clients that, on Notifiable Projects, the CDM-C has to be appointed before preparation of the initial concept design can commence
- Remind Clients that the CDM-C is then available to advise and assist with *all* of their duties on Notifiable Projects
- Advise Clients that the Designer (if competent, willing and properly resourced) or a CDM-C equivalent can be engaged to assist with Client duties on Non-Notifiable Projects if they are at all concerned that their duties could be difficult to discharge or if they are not sure how to discharge them
- Remind Clients (as well as oneself) that duties and responsibilities should be discharged in ways that are proportional to the risks on projects, that paperwork should be kept to a minimum, and bureaucracy avoided

Do Not...

- Start design work until satisfied that Clients are aware of their duties
- Start design work if the project is Notifiable and the CDM-C has not been appointed
- Offer to give advice, assist with any duties or accept an appointment for which you are not competent and capable
- Promise more than can be delivered (time scale especially)
- Frighten Clients with a seemingly endless list of onerous duties

2.9   Finally...

APS advises you not take on an appointment if a Client refuses to accept or comply with his or her duties.

Being Capable: Competence, Resources and Commitment

Remember:

Designers' duties apply from day 1 on all projects and regardless of whether or not payment is to be made

3.1 Overview

This section of the Guide covers the requirement in the Regulations for Designers to be competent if accepting an appointment and to show others, if asked, that they are capable of dealing with DRM using the skills, experience, resources and commitment that are appropriate to a specific project. While the text inevitably concentrates on issues as they relate to larger organisations, the same principles will apply to small ones and to sole practitioners. The difference will be that procedures should be simpler and the need to discuss and co-ordinate within the organisation less onerous or non-existent with the latter.

The section covers the development of appropriate attitudes, competence and resource requirements, the process of developing competence and resources, tailoring need to projects, collecting and collating information, and presenting that information to those who need to see it to show them that the capability exists to do what is required.

Being capable is, of course, more important than demonstrating it, so Health and Safety capability issues will almost always be best addressed in the context of the whole design service. Capability will also be enhanced if the Designer, or design company, can demonstrate the Health and Safety goals and objectives that are an integral part of the whole design service, including a commitment to keeping up to date with Health and Safety legislation and best practice.

The three fundamental questions that have to be answered by a design organisation are:

- Do we have the basic DRM competence and resources that our overall design service requires?
- Do we have the basic DRM competence and resources that this specific project requires?
- Can we demonstrate a commitment to delivering effective DRM in our work?

In other words, "Are we capable for this project"?

Of course, where a project is Notifiable, the CDM-C may check the competence and resources of potential appointees for the Client (where appointments have not yet been made) whereas, for Non-Notifiable Projects or where Designer appointments are made before the appointment of a CDM-C, the Client will need to check competence and resources. In both cases, however, the new regulations require that anyone accepting an appointment *has* to be competent.

3.2 Competence, Resources and Commitment to Health and Safety DRM

Looking at 'capability' can be a helpful way of assessing the suitability of a Designer or design team to integrate DRM on a project.

'Capability': 'Capability', which can be defined as Competence + Resources + *Commitment*, uses 'competence' and 'resources' as significant contributors but also takes into account the attitude to Health and Safety and to DRM of those providing the design service.

For Designers, this way of looking at competence and resources could be critical. Putting all these issues together, with an emphasis on 'concern' and 'commitment', could be the key to defining those who should be appointed on a project and enable them to demonstrate that they have the required levels of competence demanded by the CDM Regulations. Capability ought to be what Clients and their CDM-C advisers are looking for.

Competence: The first requisite of competence must be in DRM as a process, while the second must clearly be basic knowledge of construction and related Health and Safety DRM issues, including alternative approaches that can reduce risks to Health and Safety. The third requisite of competence must be in relation to the activities and structures comprising the particular project being considered.

Resources: There is then a need for appropriate resources, particularly to provide any additional input required to:

- Deal with specialist hazards or specialist design areas

- Extend Health and Safety understanding relating to construction
- Provide the necessary levels of DRM skills required to deal with project and construction types not previously undertaken

These could, of course, involve use of sub-consultants or additional staffing resources whose capability would, in turn, need to be established (*see Section 4*).

Keeping up to date: This will all be of little value, however, unless the people involved in the design team are also continually updating their knowledge and experience. A commitment to continuous professional development (CPD) and the availability of good practice information must be critical issues in assessments of 'capability'. This suggests that any 'register' or 'ticket' that is to be worthwhile in demonstrating capability/competence should include a CPD/updating requirement – like that of the APS Register of Designers, which is available to all Designers who meet an ability threshold – that provides information and guidance on project safety issues to those on it, and requires a CPD commitment.

On many occasions the design provision will be by a company and the management of the process will be dependent on the competence of the *critical people* in that company dealing with the project. The company itself can then be seen as an essential part of the 'resource' that enables the Health and Safety DRM to be effectively delivered. All these issues need to be considered when assessing capability.

Over and above all of this will be the enthusiasm, commitment and professionalism of the individual Designers (and company) that will determine the 'capability' of the company to deliver the service required.

Commitment is what can turn competence and resources into capability

3.3 Developing Commitment to Effective Health and Safety DRM

Preparing and regularly reviewing a statement of policy relating to the delivery of DRM in the work of the individual or the company could help develop a positive attitude and commitment to effective DRM. It will be useful in such a statement to include the company's Health and Safety goals and objectives and to make sure that all are aware of them.

Encouraging an understanding of the DRM process and the ability to see and explain how this has been carried through and how it has been of benefit to projects will develop a broader, positive attitude to this aspect of design, and its integration into the overall design process.

The policy on delivery of DRM will also need to be integrated with policies on CPD and personal development, and take account of developing needs to extend general and project-specific capability. This will reflect the Designer's or design company's approach to the next three elements of this section.

3.4   Core Competence Issues

It almost goes without saying that individuals must have appropriate qualifications and experience unless their work is being overseen by other competent persons, and the new ACoP has established a series of thresholds relating specifically to Health and Safety and DRM for Designers and companies to meet.

For 'Designer' companies (which will include those operating individually as Designers), the ACoP (Managing Health and Safety in Construction) sets out core criteria that cover:

- Health and Safety policy and organisation for Health and Safety[1]
- Defined responsibilities and arrangements for Health and Safety and Health and Safety management across the company
- Availability of advice on general Health and Safety matters as well as advice relating to construction Health and Safety
- Training and CPD to ensure that all in the company can effectively discharge their Designer duties (e.g. construction Health and Safety as well as CDM, DRM, etc.)

---

1   In small companies (less than 5 people) a written policy is not required – but it is difficult to see how the ACoP can be complied with, without some form of written statement

- Procedures for ensuring discharge of all Designer duties (under CDM Regulation 11) including identification and elimination, reduction and management of Health and Safety risks
- Procedures to ensure co-operation and co-ordination between Designers in a project team
- Systems for monitoring, auditing and reviewing DRM policies and procedures, (which include workforce involvement)
- Evidence of relevant work experience in relation to specific appointments
- Proportion of the company who have demonstrated knowledge/ability related to construction Health and Safety, relevant qualifications and registers, etc
- Arrangements for ensuring that sub-consultants meet these same requirements

Companies should be able to demonstrate how they meet these requirements or be able to provide certification, or show compliance, through an independent accreditation company – for example, the APS 'Directory of Registered Practices' scheme – which has been established specifically to assist with this evidential requirement.

3.5 Acquiring the Competence Needed

Competence should be developed in a planned way by:

- Identifying gaps in DRM-related knowledge
- Identifying means to fill the gaps
- Undertaking relevant study or training
- Using experienced staff, CPD programmes, or training providers to deal with specific needs
- Using on-the-job training, which may be more suited to those not yet competent who can be closely supervised by competent people

The starting point for this process will be by assessing needs against the core criteria in the ACoP (*see 3.4 above*), but could also be extended by examining the Designer Performance Standards produced by CIC. These are based on an analysis of Designer performance against the requirements of the Regulations, and tackle specific capability requirements and their implications for individual Designers in relation to their various duties (*see* www.cic.org.uk/activities/SiDCompetenceStandarde.pdf).

3.6   Capturing and Using Experience

Feedback from current and completed projects can inform future practice and should be included in project files where it may be relevant to ongoing issues. This is a normal aspect of good practice for all Designers, and feedback from Health and Safety DRM aspects of a project should be an integral part of that process. This information could also be filed ('Practice Feedback File') so that it can be called upon during future projects that involve similar types of work, site situations or risk issues.

It can be helpful in developing 'capability' to establish and develop DRM experience records by:

- Debriefing staff for experiences they have gained
- Identifying successes
- Identifying problems and contentious issues which should be avoided in future projects
- Recording experiences and conclusions
- Ensuring that everyone in the company is informed of these outcomes

Debriefing may seem time-consuming, but it can be a critical learning process for a practice and hard-won information can readily be lost without systematic feedback. Debriefing when a project is nearing completion or before staff transfer from the project to other work (or leave) can uncover and help retain useful information on DRM strategies and the successful tactics employed on projects. This will be valuable to those who have worked on the project and those who may work on similar projects at some time in the future.

Team meetings, project reviews, and peer group discussions can all be used to kick-start debriefing, and a simple checklist/questionnaire can be used to guide discussions. The fundamental need is to review project experience where the Designers have made a positive contribution to reducing risk including:

- Construction Health and Safety risk management

- Improved co-ordination and project planning
- Simplified construction
- Safer or healthier construction/maintenance
- Project programming
- Cost effectiveness

The information gained can be placed in a 'Capability File' (*see 3.8 below*) and/or a 'Practice Feedback File' that may be based on any of the following:

- Work sectors: type and nature of the work
- Building/engineering/structures types
- Size of projects
- Complexity and nature of projects

The objective, however, will be to maintain a record of positive outcomes from CDM DRM that can be easily retrieved to inform further design and practice, not to collect paperwork!

3.7 Establishing Necessary Resources

Many technical information resources covering Health and Safety in design and DRM issues are available and include:

- Internal and external libraries
- Information services – on CD-ROM
- HSE information
- Information/advice service – online/Internet/web page
- Industry sector guidance (for example, from the Construction Industry Advisory Committee (CONIAC) – an HSE committee)
- Institutional guidance on DRM, etc
- APS Guides and Practice Notes

Complementary external support may be freely available or may be available for a cost from:

- Institutes
- Associations and other specialist organisations

- Specialist consultants – firms and individuals

All of these resources will form an integral part of the overall resource capability of the design office.

Perhaps the most important resource will be the design staff working on a project or available to provide DRM assistance with specific aspects of risk management. It is therefore important to establish who will manage the DRM process in a company, especially on large complex projects; what systems are available or will be put in place to manage delivery of the DRM service; and how resources present within the organisation can be known about, accessed and made available to a specific project.

The key questions that a Client might ask are:

- Who will manage the DRM process?
- What Designers and skills are being brought to a particular project?
- How will progress information be reported?
- How frequently will DRM performance be checked and by whom (especially if there is no CDM-C on a Non-Notifiable project)?
- What QA systems will be in place?

Designers will need to be able to answer such questions.

3.8 Developing a 'Capability File'

A basic 'Capability File' covering competence, resources and commitment issues will enable a Designer or company to respond quickly to Client inquiries. Once established, such a file should be reviewed periodically and brought up to date with information covering:

- Company Health and Safety policy and arrangements[2] signed off by the company's managing director or equivalent
- DRM policy (for the whole company)

---

2 In small companies (less than five people) a written policy is not required, but it may be helpful in responding to capability/competence enquiries to be able to hand over a simple document which covers these issues.

- Health and Safety information resources and advice
- Project-specific capability
- Project experience and key staff available
- Staff changes, additional training (re. Health and Safety and DRM qualifications)
- CPD, training, policy and company commitments, especially with regard to Health and Safety, CDM and DRM competence development

The Capability File – or a suitable extract from it – will need to be supplemented with project-specific information in relation to any new project capability enquiry. Registration with an external accreditation body (such as the APS 'Directory of Registered Practices') completes this evidence of capability.

3.9 Assessing Project-Specific Capability Needs

Once core capability has been established, project-specific needs can be identified. Checking the project brief and mapping capability needs in terms of Health and Safety and DRM can help determine project-specific requirements, including those relating to:

- Client, design and contract requirements
- Construction complexity, especially where novel forms of construction may be envisaged
- Specialist installations and operations
- Specific Health and Safety hazards
- Additional CDM DRM work and advice to Clients
- Specific DRM skills and experience
- Specialist professional skills (such as knowledge of working over or under water, in deep shafts, etc)

This can be followed through by checking and listing relevant skills, knowledge and experience available to the practice based on:

- Current projects
- Project records
- Staffing information
- Peer group discussion

By examining job records to establish experience of similar work, it will be possible to establish any need to employ specialist help. Unusual aspects of the work may need to be handled by staff with the relevant knowledge – so it will be important to identify gaps that may need to be addressed. It is therefore helpful to take steps to complement existing competence and resources by:

- Identifying gaps in DRM knowledge for potential project types
- Identifying means to fill the gaps, including taking on additional staff or access to consultants
- Undertaking the study or training identified

and identifying the means of providing the necessary skills by focusing on:

- Which staff?
- What outside help?
- What additional study/training?

Finally, account should be taken of possible 'unusual' queries and expertise requirements by:

- Arranging for use of external libraries, access to HSE and APS contacts and help lines to answer unusual queries
- Arranging for liaison with experts within other (design) disciplines (in-house or external consultants)
- Using computer software to provide management support and information co-ordination

3.10 Presenting Written Information to the Client

Clear and concise documents covering the information that a Client could expect to see should include:

- Summary of the approach that the Designer or design company takes towards their work (see 3.3 above)
- Outlines of the core competence of staff in terms of general DRM training and experience
- Project-related capability, including previous relevant experience

- Details of staff, time and support resources available for this project
- Any additional information that is needed to cover specific Client and project requirements
- Information on other relevant matters such as Professional Indemnity Insurance (PII) cover, backup in case of accident or ill health to critical staff
- A note to make it clear that further information is available if required
- Accreditation Certificate, if available

It may, of course, also be necessary to provide evidence of competence from records and samples of work in addition to prepared documents if requested by a Client to do so.

3.11 Presenting Information to the Client Verbally

Preparing a short presentation to use at an interview or meeting makes sense – but only if carefully thought through and rehearsed so that time and content are appropriate and project-specific. This should be based on material collected together for written presentations, and it may be helpful to leave behind a précis of the presentation and/or a copy of the full written evidence if this has not previously been submitted. If it has, the presentation will probably need to focus on additional project-specific issues or explore issues in more detail than the written presentation.

3.12 Warning

Designers accepting an appointment are required, under the CDM regulations, to be competent. If, after assessment of project-specific competence and resource issues, it does not seem possible to plug any identified gaps, the most sensible response will be to walk away. There is no disgrace in concluding that your organisation does not have the capability to fulfil the needs of a particular project – it is the mark of a professional to know his or her limitations.

Capability of Other Designers

4.1   **Capability of Sub-consultants and Others**

As everybody involved in designing for construction work has to be competent, it is obvious that this must include all design sub-consultants and others who may be engaged or appointed by Designers to assist them with their work. This applies to sub-consultants who provide additional resources or work on specific aspects of the commissioning Designer's work, as well as those who carry out other types of design work that is outwith the appointing Designer's competence. So, where work is sub-contracted or other appointments are made by a Designer for design work to be carried out by others (not part of the same company), reasonable checks on the competence and resources of those providing that service will have to be made.

However, a commitment to DRM and Health and Safety management in design will often be as important as a demonstration by the sub-consultant/consultancy that core competence criteria have been met. What Designers also need from their sub-consultants is 'capability'.

Competence under the CDM Regulations means Designers should:

- Understand the construction processes that are required to fulfil their design
- Be able to identify the hazards associated with such processes
- Recognise any significant implications for Health and Safety associated with those hazards

They should then be able to:

- Modify their designs to mitigate risks by eliminating, reducing or controlling them
- Identify significant risks which remain in construction, use, maintenance or demolition and pass on that information to those who need it
- Demonstrate that there is at least one way to construct their design safely
- Properly consider workplace Health and Safety issues in their sub-consultancy design work
- Co-operate and be willing to co-operate with others to enable project DRM to be effective

The processes and criteria for assessing the competence of sub-consultants will be the same as those set out in Section 3 for assessing the competence of any individual or Designer company. Sub-consultants should be able to show that they can conform with these. If a design element is further sub-contracted, it may be wise to check on the competence of any (sub)-sub-consultant.

4.2   Resources

'If you deal with the lowest bidder, it is well to add something for the risk you run ... and if you do that, you will have enough to pay for something better'

John Ruskin

It is against the law to allow a Designer to start work until those making the appointment have satisfied themselves that the job will be properly resourced.

For the purposes of the CDM Regulations, 'resources' fundamentally means 'time and money'. For a Designer who is engaged as a sub-Designer or sub-consultant, this means that the fee for the job must be appropriate and that adequate time is being allowed for production of the designs. It is the responsibility of those making the appointment, or arranging for that work to be done by someone else, to ensure that both of these are adequate for the task in hand.

If either the sub-consultant's fee or time is inadequate, it is likely that the product will be inadequate, i.e, the design will not satisfy the requirements placed on Designers by the CDM Regulations. As this situation is against the law, it is likely that the appointing Designers will have to make good this inadequacy, most probably at their own cost, so negating any time or cost saving.

4.3   Sub-consultants Located Outside the UK

Remember that the requirements for competence and resources apply to all sub-Designers or sub-consultants, and this will apply even when they are located outside Great Britain. The commissioning Designers will therefore need to carry out the capability checks referred to above so that they can be confident that the designs of sub-consultants can be built, maintained, used safely (if a workplace) and demolished.

Otherwise they will need to check that all of their sub-consultant's work has been carried out in accordance with the CDM Regulations and that DRM has been effectively applied at all stages. This could lead to modifications of the sub-consultant's designs with all the consequences that follow – not least that these designs may no longer be appropriate in other respects. The sub-consultants may also refuse to accept any liability for designs that have been changed.

It would therefore seem to be both prudent and more effective in the management of DRM to make sure that those appointed or engaged are capable and have adequate resources available to them. Notwithstanding this, for any designs procured overseas it will remain the responsibility of the UK-based Designers (as procurer or commissioner of any sub-designs) to ensure that the designs meet the requirements of the CDM Regulations.

4.4   Novation of Sub-consultants

The fact that a sub-consultant has been appointed previously should mean that competence and resource checks have been carried out in relation to the designs that they were expected to prepare up to the point of novation – and it may be the case that these checks have covered the further design work required.

However, before accepting a novation, it would be prudent for a Designer to check the competence and resources of the sub-Designers or sub-consultants, as the necessary checks may not have been carried out or may not have covered work that will now be required under the new arrangements. If, as a result, the sub-Designers or sub-consultants are not deemed capable, the novation should be refused.

4.5   Co-ordination of Sub-consultant's Work

All the regulations that apply to Designers also apply to sub-Designers or sub-consultants. It is therefore important that they are treated as design team members, either directly or indirectly. A commissioning Designer has to decide how to incorporate their input and ensure the flow of information in both directions. Treating sub-Designers or sub-consultants as part of the commissioning organisation and integrating them into the overall design co-ordination procedures will ensure that their contributions to DRM are effective.

4.6 Records and Feedback

When it comes to defending an allegation that a design is deficient, it will always be a good defence to be able to demonstrate that reasonable professional judgement has been exercised. For the purposes of the CDM Regulations, this means that any decisions taken should have been based on as full a consideration as practicable of the Health and Safety issues associated with an item of work. This will include the competence and resources of those appointed as sub-Designers or sub-consultants.

The records of the checks made prior to any appointment or arrangement for work to be carried out by others could therefore play an important role in demonstrating this.

Records of the enquiries made, bearing in mind that these only need to be 'reasonable' in relation to the work to be carried out, should cover the essential elements of competence and resources assessment outlined above (and in Section 3) together with a summation of the responses made to these enquiries. The record should then include the assessment and judgements that resulted from this process, and should demonstrate that reasonable professional judgement has been exercised in relation to these sub-Designer or sub-consultant appointments, engagements or arrangements for work to be carried out.

4.7 When Things Go Wrong

If a sub-consultant is not competent, it is likely that their designs will be inadequate because they will not have given adequate regard to Health and Safety matters. The design will not satisfy the requirements of the CDM Regulations and may not be safe to construct. Such designs will therefore have to be made adequate, perhaps resulting in repeated work for the sub-Designer or additional work by the commissioning Designers. As this will cost time and money and impact on project progress, the need to carry out the competence checks and assessments before appointments are made is clear.

In addition, the shortcomings of a 'less than competent' sub-consultant's design could result in other design decisions being made that include unnecessary or unacceptable risk which could lead to ill health, injury or death. This could readily be the case where finalised designs depend on the sub-consultant's designs, and shortcomings in these have not been recognised or identified.

If misfortune strikes and a sub-consultant is found to be less than competent, prosecution for appointing an incompetent person could well follow – and this will be difficult to defend if checks have not been carried out or if records have not been kept. It can be difficult to recall events that took place a considerable time before the resulting prosecution, and verbal evidence will not be as reliable as written records.

The ACoP makes clear the need to eliminate unnecessary paperwork in CDM processes, and there is no requirement in the regulations to keep a record of competence checks of sub-consultants (prior to appointment). However, it might be difficult to establish that competence checks have been carried out unless some form of record is kept. These need only be simple records – and they should be proportionate and reasonable in relation to the issues – but records should be made of the issues checked, the assessment made and the conclusions drawn.

4.8 Overall

Do…

- Check the competence of any sub-consultants who may be commissioned
- Check that the sub-consultants actually have the resources to deliver their duties and for designs to be completed
- Check that the sub-consultants have systems in place to help them to comply with their duties under CDM
- Set up a communications and co-ordination system with the sub-consultants that is auditable
- Allow time and fees for the sub-consultants to attend design review meetings

Do not…

- Make the process of checking competence disproportionate
- Accept novated sub-consultants who cannot demonstrate their competence
- Assume adequate resources will be available to the sub-consultants
- Skimp on the fees that you pay

Remember:

- Less than competent sub-consultants could lead to the prosecution of the commissioning Designer for the sub-consultant's inadequacies
- Less than competent sub-consultants could make it extremely difficult for the commissioning Designer to comply with their CDM duties
- Inadequately resourced sub-consultants are likely to be late with information, which could lead to the commissioning Designer giving hurried and incomplete consideration to important CDM matters
- Less than competent or under-resourced appointees are likely to provide inadequate designs and could cost the commissioning Designer money

Co-operation and Co-ordination

5.1   Introduction: Co-operation and Co-ordination

Although in common use and familiar to all, these two words have now been given specific importance for Designers working under the new CDM Regulations. It might therefore be helpful to consider the dictionary and thesaurus definitions of these terms:

Co-operation	Co-ordination
Is: Working as a team to achieve common aims; Combining efforts; Assisting each other; Being cohesive and Collaborative	*Is a process of: Organising; Directing; Managing; Synchronising; Harmonising; Matching up and Bringing together*

The context for any of these interpretations is, of course, simply that of making designs work well, making them safe or safer, healthy or healthier and ensuring that they perform optimally together in a developing or completed structure.

In simple terms, 'co-ordination' requires that all the parties involved provide or do something in an integrated way to an arranged timetable, while 'co-operation' is the obligation to work with others to do this. Under the CDM Regulations, Designers are obliged to co-ordinate their activities with each other, to co-operate with others and to seek the co-operation of others.

5.2   Integration of All Co-operation and Co-ordination on Projects

One essential aspect of Health and Safety-related project co-ordination (and co-operation) is that it should NOT be something that is separate from all the other project co-ordination that takes place, or should take place, on projects.

Co-ordination to secure the Health and Safety of those working on, or affected by, construction work will inevitably have an effect on overall project co-ordination. So it will be the case that, on some projects, this requirement to co-ordinate will encourage or stimulate other co-ordination – or even result in overall co-ordination taking place where it might not have done so. This is therefore one of the big changes that the new

CDM Regulations have wrought – the *requirement* for work on projects to be co-ordinated.

The integration of Health and Safety-related project co-ordination with normal design processes will also help ensure that DRM has only a light effect on overall project procedures and systems (assuming that they would be there in the first place) and documentation and processes should be developed, not invented, to enable effective co-ordination and co-operation in relation to Health and Safety issues. In this way, unnecessary bureaucracy can be eliminated and the process made 'results focused' and effective.

5.3 Co-ordination: the Key to Health and Safety Risk Management

If DRM can be said to have two legs, then 'co-ordination' and 'information' are these legs – and the latter without the former is almost pointless. Co-ordination is therefore the key to effective DRM.

On Notifiable Projects the CDM-C is required to ensure that suitable arrangements are made and implemented so that co-operation and co-ordination can take place and arrangements will be made in conjunction with or by the CDM-C. However, on **Non-Notifiable Projects**, where there is still a requirement for all involved to co-ordinate their activities with each other, there is no-one charged with making sure that this happens other than the Client. On these projects it is the Client who has to ensure that arrangements are made for managing the projects. This is a duty that will present a challenge to many Clients who may expect other members of the design or construction team to assist with, or at least advise them on, these co-ordination management issues. Designers will therefore need to understand what is practical and effective to achieve these objectives.

5.4 The Focus of Co-operation

The principal objective of project co-operation will always be to ensure that designs are compatible in terms of Health and Safety, as well as being suitable in themselves. Because Designers will always be considering construction, maintenance, demolition and workplace Health and Safety issues alongside other requirements and influences, there will be an overriding need to co-operate with all Designers at all stages of a project. For instance, a Designer may not be aware that an element of their design will create a hazard when taken into consideration with an element designed by another. Once this has been recognised, it may be possible for both Designers, working together, to generate a solution that avoids or minimises this situation. (A simple example might be where an architect for aesthetic reasons designs an element that creates construction safety difficulties for the structural engineer attempting to design a support system).

Every Designer will need to know of any risks generated or remaining in one aspect of a design so that they can be taken into account when other aspects of that design are being developed. For this reason, residual risk information will need to be made available to others throughout and across the project. This risk information will include:

- Hidden or unusual hazards or risks associated with site conditions
- Novel or technically challenging solutions
- Construction risks that would not be obvious to a competent Contractor or Designer
- Constructions that are likely to be difficult to manage effectively
- Significant risks arising from any aspect of the design

5.5   Notifiable and Non-Notifiable Projects?

Co-ordination and co-operation on Notifiable Projects: On all projects, Designers should identify risks on a continuing basis (not just at the end of a work stage) and co-ordinate their work with that of others in the design team. However, this co-ordination process will be a key aspect of the CDM-C's role in the project team on Notifiable Projects. The CDM-C will, as previously noted, ensure that suitable co-ordination arrangements are made and will work with others to identify appropriate co-ordination processes and procedures.

All Designers will have to co-operate with others to ensure that their designs are compatible with other designs and will need to take whatever steps are appropriate to resolve any Health and Safety risks or conflicts. A key step will clearly be the production of project design Health and Safety information to be passed by (through, or with the knowledge of) the CDM-C to those who need it – whether for further design work or for construction work. The CDM-C will also work with Designers to ensure that the process of risk reduction and management is effective, and Designers will obviously need to co-operate with the CDM-C to achieve this.

Co-ordination and co-operation on Non-Notifiable Projects: On these projects, where there is no requirement for a CDM-C to be appointed under the CDM Regulations, there will nevertheless need to be someone who will take the lead and make sure that the various aspects of the work are co-ordinated. Whilst every Designer is required to co-ordinate their work with that of others nevertheless there will need to be someone who understands the overall project circumstances and inherent risk issues and can:

- Identify the type of work involved in relation to the layout and general environment of the site and identify inherent risk issues
- Identify detailed Health and Safety issues themselves and those raised by Clients, other Designers, Project Managers and other consultants
- Establish inherent design risks from project concept information or outline design
- Establish what additional information may be needed at the outset and throughout the project

Designers and Clients may need to consider the selection of a lead Designer who can help the Client ensure that management arrangements are put in place (this may also be useful on more complex Notifiable Projects). The lead Designer can then act as a focus for project co-ordination and co-operation and enable:

- Effective co-ordination of design work
- Systematic, collaborative and compatible approaches to hazard identification, etc
- Consistent DRM priorities and methods
- Free flow of information between Designers

On more complex Non-Notifiable Projects that involve significant inherent risks, the Client could appoint a 'quasi CDM-C' to oversee co-ordination and co-operation and assist the Client with other duties.

5.6 Critical Issues for Effective Co-ordination and Co-operation

Commitment: Co-operation should take place at every contact – telephone call, email, meeting and other forms of dialogue and exchange – between team members. But co-operation does not just happen! Designers need to make positive decisions about their engagement in the co-ordination and co-operation processes and need to participate in a continuing dialogue to ensure that Health and Safety or welfare issues for construction work are not overlooked or underplayed, especially on Non-Notifiable Projects.

Resources: Co-operating and co-ordinating should be a normal aspect of design work but, where other services to the Client are provided on Non-Notifiable Projects (such as lead Designer, Contract Administrator or Project Manager), it will be worth checking that the resources needed to provide design Health and Safety co-ordination, co-operation and leadership are available.

Identifying Designers: In conventionally procured projects it is usually straightforward to identify the Designers. However, this may be more difficult under some procurement methods as these may change at different stages of a project, and because Designers may be an embedded part of Principal Contractor operations and will often be working for specialist Contractors. On all projects (including Non-Notifiable Projects), the Designers will have to identify other CDM Designers and will need to make contact with relevant Designers, including those involved in specialist work packages and temporary works, to ensure co-operation and enable co-ordination. This process will, of course, continue throughout the construction phase.

Working together: Design team members need to have knowledge of the ways in which others in the team are identifying hazards and an understanding of what they consider to be of 'significant' risk. The use of common methods can help, and it may be useful for a lead Designer on a Non-Notifiable Project to offer to co-ordinate this aspect.

Co-ordination methods: A simple list of useful methods for developing effective co-ordination and co-operation, as well as input to the Health and Safety issues in the project, would include:

- Setting up an integrated team involving Client, Designers, CDM-C (on Notifiable Projects), Principal Contractor and other relevant Contractors
- Appointment of a lead Designer where many Designers are involved
- Placing Health and Safety in design issues as a specified agenda item on all project meetings
- Holding a 'brain-storming' Project Risk Management Workshop for the whole project team early in the project to identify inherent risks and to decide who will be dealing with them, including Health and Safety related risks
- Agreeing common approaches to DRM and how to record it
- Using design review meetings to deal with DRM issues throughout design development stages
- Carrying out critical stage Health and Safety hazard and risk reviews by the whole project team as each design stage is completed
- Holding regular meetings of all the design team (including the CDM-C) with Contractors and others
- Using a tracking system – such as a hazard and risk register – to maintain and develop a chronological log of fixed and changing issues

These methods are, of course, equally applicable to Non-Notifiable Projects as to Notifiable Projects – but in the former case, because there is no CDM-C, someone has to make sure that they happen. Under the CDM Regulations this is clearly the Client but, for all practical purposes, it will usually be someone appointed to do so *by* the Client. Because this additional role and service will demand skill and resources, those taking it on will need to consider the competence and fee implications. On short projects the relative cost could be significant and the level of knowledge and cross-discipline experience required may not be readily available. These are not issues that should be overlooked.

Co-operation and co-ordination with others: There is also a need to look beyond the immediate project issues and to consider the possible impact of the construction work on those using adjoining land, on adjoining construction work by others and, in turn, on the possible impacts that other work on the site or on adjoining sites could have on the proposed project. Co-ordination and co-operation will need to consider such issues, and liaison with those working on other sites will be a necessary starting point. Clearly 'adjoining' in this context means any site nearby, not just those with a common boundary with this project.

Workshops: To many in the construction industry, the terms 'Risk management' and 'Risk Management Workshops' are usually interpreted as being concerned with 'whole project risks' – and will cover every type of risk that may affect the completion targets of a project. While some very significant Health and Safety issues may appear on a project risk register, they will usually be just a few issues among many non-Health and Safety-related project risks. However, if Health and Safety issues are considered alongside other project/construction risks, it is more likely that major players (for example, the Client, project manager, design team principals and leaders, facility managers, maintenance Contractors, etc) will be encouraged to get involved and stay involved. It should therefore be recognised that workshops concerned with Health and Safety risks and the related Heath and Safety Risk Management are a subset of whole project risk management and can make a significant contribution to this.

A major benefit of an early Health and Safety Hazard and Risk Workshop will be that project risks can be identified in a holistic way and the whole team made aware of significant risk issues relating to specific aspects of design. In addition, the workshop can be used to agree co-ordination methods and encourage the allocation of responsibility for each aspect of risk so that it is clear who should be dealing with what. Such workshops can be useful throughout the design process but are particularly important at its outset. On Notifiable Projects, the CDM-C will have a significant role in these workshops – yet another reason to make sure that the CDM-C is in place before design work commences.

Note: While the CDM Regulations do not directly require Designers and Contractors to provide the Client with Health and Safety File-related information from subsequent Non-Notifiable Projects, the fact that a Health and Safety File compiled under CDM 2007 has to be subsequently kept up to date by the Client implies that the Client needs to be provided with relevant information from Non-Notifiable Projects. In the future, Health and Safety Files should be a comprehensive source of existing risk information.

5.7 Designs Produced Outside the UK

Under CDM 2007, all duty holders have to co-operate and co-ordinate over project Health and Safety. In an integrated project team, this should take place as a matter of course. However, where, for any reason, designs are produced by Designers from outside Great Britain, the need for those designs to be suitable and compatible with any other project designs from within Great Britain remains. Designers from outside Great Britain have no duties under CDM 2007 but, when these designs enter Great Britain – typically through the lead Designer or the procuring practice or Contractor – then these organisations have the legal duty to ensure they meet the Health and Safety requirements of Regulation 11. Effectively, this invokes the wider duties of Designers to co-ordinate and co-operate.

So, in simple terms:

- Design decisions made outside Great Britain have to meet CDM 2007 standards and Designers in Great Britain have to ensure that they do so when they adopt, adapt or simply pass on these designs
- Design practices in Great Britain have to have in place a review process to ensure that designs procured from outside Great Britain have responded appropriately to CDM Regulations, carried out effective DRM and provided appropriate residual risk information

Designers who source work abroad are therefore faced with two options – both of which necessitate resources, planning and management:

- Provide non-British Designers with information on DRM principles, a brief and a system for informing the UK commissioning Designer of any residual risks and any information that should flow from them to the supply chain *Or*
- Provide the non-British Designers only with a design brief and review on receipt to identify any residual, significant or difficult-to-manage design-based Health and Safety risks, and then take steps to apply Designer duties under Regulation 11

This second option is definitely not the preferred option because a key principle in the duty of Designers is to eliminate hazards and reduce risks where the hazard remains. This is most effectively done as the design proceeds, and to wait for information to be issued could be to waste opportunities or result in abortive design work.

In either case, it would seem that those who commission outsourced designs will need to establish some mechanism or process for ensuring that these designs can be co-ordinated effectively with the designs of others in the project team – the difficulties of which may be exacerbated if those others also include Designers working abroad.

5.8   The End Product

Co-ordination: The outcome of effective co-ordination will be seen as good design that provides the necessary solutions and controls of the specific risks relating to a specific construction, its maintenance and demolition.

To the question: '*Is this design solution unusual or complex or does it still include health or safety risks to those having to work on creating or maintaining the physical solution?*' there will be one of two answers:

If the answer is 'no', then it is likely that the Designer will need to do no more than to communicate this fact. However, it will be essential to communicate the fact that there is nothing significant – so that others do not waste time looking for information that has not been provided.

If the answer is 'yes' or 'depends on another design input', then the process of co-ordination must continue.

Communication is the key. The outcome of all DRM should be designs that are capable of being constructed or maintained safely and with minimal health risks. Once this has been achieved, relevant information must be delivered to those in the supply chain who need it. This will enable others to co-ordinate their activities with Health and Safety in mind.

Range of possible information outputs: There is a range of information outputs that a Designer will generate and the use of items in that range will reflect design practice preferences and style. The range of possible information outputs includes:

- Normal specification or production drawing information
- Records of residual hazards or risks in project records or a co-ordinated project hazard and risk register managed by (typically) the lead Designer on Non-Notifiable Projects and the lead Designer or perhaps the CDM-C on Notifiable Projects
- More detailed information issued with designs covering safe sequences of construction, likely solutions or specified methods of construction
- Specific information on critical site or structural conditions that can only be co-ordinated effectively by passing on detailed information in survey, investigation or inspection reports

The extent of each of these outputs will, of course, be proportionate to the risks remaining.

Co-operation: This is about the exchange of information and ideas, a willingness to adapt elements of one's own design to complement those of others and also offer constructive comments and suggestions to others to help improve the whole design. Co-operation is about the style or approach to working with others.

To understand the relevance of co-operation, we have to understand the interactive implications of the work of others on designs, as well as the impacts of those designs on end users and those who construct or maintain them. As with co-ordination, the outcome needed from effective project co-operation is to produce designs that are capable of being constructed or maintained safely and with minimal health risks. This must be judged within the context of holistic solutions that meet the Client's brief and satisfy cost, time and quality targets. In other words, co-operation – like co-ordination – is an integral element of effective design.

Information to Inform Design and Planning

6.1 Setting the Scene

One of the critical elements of the CDM process is that of ensuring that 'the right information goes to the right people at the right times'. This is fundamental to co-ordination and co-operation and it is one of the essential tools to enable effective DRM. Without it, those working on designs may be unaware of many interactions that can create risks; without it, those working on projects may remain exposed to residual risks.

Get the Right Information to the Right people at the Right Times

Getting the right information to the right people at the right times is therefore a major objective in the DRM process. However, DRM is not about passing around information but about reducing risks to those working in construction – so making sure that relevant information goes to those who need it. The main determinant of the extent of this information flow is to enable risks to be reduced.

To be able to make sure that relevant information goes to those who need it, there must first be an understanding of what is and what is not 'relevant' information. All available project information must therefore be identified, collated, assessed for its relevance to possible risks, and the relevant risk-related information passed on to those who need it. Critical to this process is ensuring:

- That information is seen to be relevant
- That someone is taking an overview of what may be relevant to all those involved in design and management planning
- That someone is making sure that they receive that information at the time they need it

These are critical duties of the CDM-C under the new regulations, and Designers will need to recognise this fundamental aspect of information management on Notifiable Projects. On Non-Notifiable Projects, Designers may need to manage this flow of information themselves, unless the Client undertakes to do this or appoints someone to deal with it.

This section of the Guide therefore discusses the effective flow of appropriate information, gives guidance on the best means of identifying it, and then suggests how the required distribution can be achieved so that Designers can in turn discharge their risk elimination, reduction and management duties under the CDM Regulations.

This section deals with all types of projects, whether new-build construction, extensions, refurbishment, services installations/alterations or demolition projects but, of course, the type of project will influence choice of the most effective ways of seeking and distributing information.

6.2 Is the Project Notifiable or Non-Notifiable? How Does That Affect Information?

The CDM Regulations impose specific requirements regarding information for both Notifiable and Non-Notifiable Projects. The situation is clear-cut for Notifiable Projects where the Client's duties in relation to management arrangements and information (Regulations 9 and 10) are undertaken by the CDM-C. However, for Non-Notifiable Projects, these duties still have to be discharged without the Client having anyone, under the regulations, to assist.

Clients may well expect Designers to provide guidance on how their management and information provision duties can be discharged. This would involve the Designer in guiding the Client on what information might be needed for the project, how it could be obtained and distributed, as well, perhaps, as assisting with those processes.

In addition, the Designer's duty to make 'sufficient' information available with any design (information that others might need to be able to comply with *their* duties) implies that Designers need to provide all such information in their possession that may be relevant to other Designers. This places the Designer in a more critical position regarding information flow as opposed to information use.

Clients will probably expect Designers to assist with these matters. On simple projects this may not be onerous, although it might attract an additional fee for the service to the Client. However, on complex intensive projects with significant risks, information needs and flow can be considerable and someone may need to carry out, in effect, the same work that the CDM-C carries out on Notifiable Projects. This could place Designers in the position of having to exercise skills and knowledge that exceed their existing

Non-Notifiable Project (NNP)	Notifiable Project (NP)
• No legal requirement to appoint CDM-C therefore: Client needs to decide the following and ensure the right information is circulated, e.g. • Who will identify what information is available? • Who will identify what information is required to be provided? • Who will identify where the information is to be distributed? • Does each party know or realise what information they require? *Clients may well 'expect' Designers to assist with all of this but on certain projects it might be of benefit to all parties to appoint a 'quasi CDM-C'. It should not cost a great deal but Clients would be more certain that their duties are being properly discharged.*	• CDM-C has to be appointed • CDM-C has to ensure that co-ordination measures are in place • Information will be channelled to, through or by the CDM-C to the various parties • CDM-C will promptly issue appropriate information to various parties

competences and resources, in which case it might be sensible to advise the Client to appoint someone separate to assist with these duties. There will, no doubt, be some CDM-Cs who would be happy to provide this additional service on Non-Notifiable Projects.

6.3 Project Information Resource and Information Flow

All projects accumulate, generate and use information that relates to and informs design, planning and works on site. Such information is an essential and critical resource in any project. For the purposes of the Guide, the APS labels this information as the Project Information Resource.

This Project Information Resource can be located in one place, as if a library, or distributed across a number of locations – for instance, it can be held by various

Designers, sub-Designers, stakeholders, Contractors, and the Client. It may be subject to an overall referencing structure (as it would be in an extranet) or it could be structured according to the systems of those who hold the information in a distributed system (e.g. indexed working papers and reports, drawing registers and file referencing systems in each office). The Project Information Resource will change and grow as a project develops to include:

- Basic information provided by a Client at the earliest stages of a project to inform the early stages of design and planning
- More robust information that informs more detailed design work (e.g. information obtained from previous Health and Safety Files and by enquiries, investigations and surveys)
- Information generated by teams of Designers as the project progresses
- Construction records, as-built drawings, building/project/Operation and Maintenance (O & M) manuals and Health and Safety File(s) handed over at any milestones in the project and at the end of the construction phase

On Notifiable Projects the CDM-C will, in most instances, instigate development of the Project Information Resource and the Hazard and Risk Register within it (*see below*); on Non-Notifiable Projects someone else will need to do this – probably the lead Designer.

Information and Notifiable Projects: On these projects the CDM-C has a duty to provide promptly relevant Health and Safety-related information (called 'pre-construction information' under the Regulations) to every person designing the structure(s) on the project, and to every current or future Client-appointed Contractor, so that suitable and compatible designs can be developed, constructed, maintained and eventually demolished safely. The CDM-C also has to prepare or update the information necessary for the Health and Safety File(s).

The CDM-C will therefore oversee the continuous provision of Health and Safety information drawn from the Project Information Resource (*see Figure 1*) as is appropriate to meet different needs at different times throughout the development of the project. In this way, the CDM-C ensures that the right information reaches the right people at the right times.

Figure 1 **Project Information Resource (PIR)**

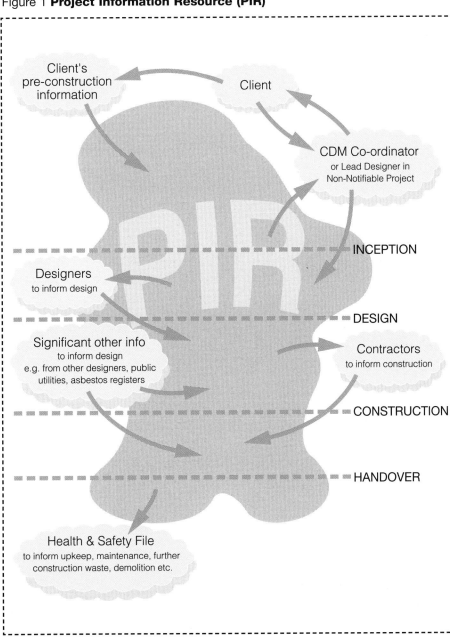

Information and Non-Notifiable Projects: On Non-Notifiable Projects the Client has the responsibility of making sure that information goes to those who need it and may need assistance from Designers with a number of these issues.

Packages of information: The information may take the form of a range of packages, bundles or a few simple documents containing the necessary information. If so, each of these packages will be tailored to meet the needs of different people at different times and for different purposes – for example, **Design Information Packages, Construction Information Packages** and packages of information relating to maintenance, repair or demolition, etc which will, of course, be part of the Health and Safety File(s). The most important of these packages in terms of project targets for construction will be the package that is provided to the Principal Contractor or those competing to be a Principal Contractor – the **Pre-Construction Information Package (PIP).** In addition, there will of course be the **Health and Safety File** package at the end of the project.

> *Packages do not have to be large bundles of paper – they could be electronic packages, access to a library, packages on an intranet resource – or, on simple projects, a few brief notes*

Who knows what is needed for whom? The right information must be made available to those who need it, even though they may not *know* that they need it, and even when others who may have produced that information do not realise that someone else needs it. On Non-Notifiable Projects where there is no CDM-C appointed, *someone* will need to assess the relevance of information to various parties; this function may fall to the lead Designer. On Notifiable Projects, facilitating co-operation and co-ordination is one of the essential duties of the CDM-C.

Information flow: The project team (including the CDM-C and, if necessary, facilitated by the CDM-C) will have to put in place processes and procedures to ensure that this flow and provision of information and co-ordination happens (for example, by design review meetings, risk management workshops, drawings and schedules issue procedures, extranet systems, etc). However, this must be proportionate to the nature of the project and the information that goes with it.

To, through or known to the CDM-C: It would also appear that the new regulations require that *all* information that may have *any* relevance to Health and Safety will have to pass to, or through, the CDM-C or, at the very least, be brought to the CDM-C's attention on Notifiable Projects so that its relevance to Health and Safety can be assessed and so that it can be passed or directed to those who may need it or may need to take account of it. This suggests that the CDM-C has to have knowledge of almost all the project information – if only to be able to be sure of what does, or does not, have relevance somewhere or somehow to Health and Safety risk management.

This does not mean that the CDM-C needs to handle every piece of information, but it does mean that the CDM-C needs to be fully aware of the extent and nature of project information as it is generated, becomes available, or is developed.

Keeping track of changes: Information relating to the hazards and risks on a project may change as designs develop and change. These changes will need to be tracked so that the implications of design changes can be passed on to those who need that information to ensure that the resultant designs are compatible and suitable. One simple way of tracking and managing Health and Safety hazards and risks, and the actions taken by Designers, is by the use of a Hazards and Risk Register. This can be tailored to the needs of the project (paper, spreadsheet or extranet-based) and will then be a primary source for the generation of the appropriate Design, Construction or Pre-Construction Information packages. An advantage of the Hazards and Risk Register is that it will only include information that Designers and the CDM-C have established is, or could be, relevant to Health and Safety.

The Hazards and Risk Register will not replace the Project Information Resource but is part of it and, in many ways, could be like an index – cross referencing and summarising relevant Health and Safety information and tracking changes that take place. Designers may use or be provided with information from the Hazards and Risk Register and will contribute information to it and the Project Information Resource, especially the residual risk information that other Designers may need so that they can co-ordinate their designs effectively.

Pre-Construction Information Package: One major focus for Health and Safety information will be the Pre-Construction Information Package. This aims to provide a level playing field when tendering, negotiating or arranging quotations for construction work. Significant elements of this will be contributed through the DRM process undertaken by each Designer. The Project Information Resource therefore clearly needs to include all the information that will be needed to compile this package – including Client requirements regarding access and the continuing use of existing or adjacent facilities.

Someone has to be responsible for providing this Pre-Construction Information Package using information in the project Hazard and Risk Register. On Non-Notifiable Projects this will probably be the lead Designer, while on Notifiable Projects it will probably be the CDM-C.

On a simple project the Project Information Resource consisted of a set of drawings and two sheets of risk related information – a simple Hazard and Risk register...

On a more complex project the Project Information Resource was based on an intranet service that all members of the team could access. This enabled the CDM-C to understand what information might need to go to whom, and when, and where additional risk related information might be needed – all for discussion at the next team meeting

Proportionate size and complexity: Remember, however, that a Project Information Resource and its related Hazards and Risk Register will be simple and brief where the information needs of a project are similarly limited while, on a complex and lengthy project, information needs could be extensive and its organisation would need to reflect this. Above all, the information resource and procedures need to be appropriate and proportionate to projects and risks!

6.4 Identifying Sources of Information

There are various sources of risk-related information that will help ensure that Designers can carry out their duties properly and fulfil their obligations under the CDM Regulations. Simple starting points for gathering that information are:

- Client information and existing Health and Safety Files

- 'As Built' drawings
- Existing risk registers
- Existing or commissioned surveys reports covering:
  - Ground investigations
  - Contamination
  - Asbestos
  - Existing services
  - Structural conditions (including for demolitions)
  - Topographical surveys (showing, for example, existing drainage, overhead services, underground storage tanks, etc)
- CDM-C on Notifiable Projects

The project information needed should, of course, take account of design, programme and contract requirements and the information needed is to establish what risks exist, especially those that need to be taken into account in the early stages of design. Knowing who to ask, what to ask and where to obtain information is the crucial issue to discuss with the Client at the earliest time – given that the Client has the duty to promptly provide pre-construction information to Designers.

6.5   Obtaining Information

(a) Client information: Under Regulation 10 the Client has the express duty to ensure that all the information in his possession is promptly provided to Designers and the Principal Contractor and other relevant parties (via the CDM-C on Notifiable Projects).

This would include early provision of information from an existing Health and Safety File as well as other available information.

Designers on Non-Notifiable Projects might ask:

- Is there an existing Health and Safety File?
- Does the Client have any information on residual risks?
- Is there an Asbestos Register?
- Are there any 'As Built' drawings for existing structures?
- What survey data is already available?

- What surveys now need to be commissioned by the Client?

On Notifiable Projects, critical concerns for Designers will be:

- Has the CDM-C been appointed? If not, design work should not start until that appointment has been made
- Once appointed, the CDM-C will co-ordinate or make sure that co-ordination takes place, and receive and provide information that is needed (or see that it is provided) as soon as it becomes available
- To work with the CDM-C and the Client to identify additional information that is needed
- To make sure that any early stage design takes account of existing site-related risks and those that existing structures present or could generate
- To wait for critical risk information if it is not available so that designs can take account of such risks

It is at the earliest stages of design that DRM can be the most effective, provided the necessary information is available

(b) Existing Health and Safety File: If the project involves working within, or extending, or demolishing, an existing structure and the existing premises were constructed around or after 1995, there should be a Health and Safety File and this should be held by the Client.

Note: Whilst the CDM Regulations require Designers to provide the Client with Health and Safety File-related information for any design, whether Notifiable or Non-Notifiable there is no such duty on a Contractor for Non-Notifiable projects. The fact that a Health and Safety File compiled under CDM 2007 has to be subsequently kept up to date by the Client implies that the Client needs to be provided with relevant information from Non-Notifiable Projects by Contractors as well as Designers. In the future, Health and Safety Files should be a comprehensive source of existing risk information.

Where is it held? The Client will have been advised of his duty by the Planning Supervisor (or CDM-C under the new regulations) to ensure that the Health and Safety File is kept available for use of all duty holders who will be working on or within the existing structure or in the curtilage of the site; it should be available, it should be on the premises and, if the Client has been properly advised, it should now be being kept up to date.

The APS 'Guide to CDM Co-ordination' provides further and fuller information on the duties of the CDM-C. This may be of assistance to a Designer who needs to help a Client with his duties relating to information provision (and co-ordination) on Non-Notifiable Projects.

(c) Information on existing risks: On Non-CDM projects in the past and on Non-Notifiable Projects in the future, residual risk information may have been retained by Designers or Clients in risk registers or other project files as there was no 'formal' location for it. Nevertheless, this information could relate to such issues as existing asbestos-containing materials that have been located but not removed, unusual structural elements or specialised window cleaning or high level maintenance requirements that could have a knock-on effect on future construction work safety.

A request to have existing risk registers or old risk-related files concerning the structure or site made available could therefore open up avenues to significant useful information.

(d) 'As Built' drawings: Similarly, any existing 'As Built' drawings which may identify Health and Safety implications relative to a project (e.g. significant eaves projections or overhangs, high level maintenance requirements or other detailed conditions with Health and Safety implications) must be made available to the whole design team and the Contractor on Non-Notifiable Projects at the appropriate time so that information can be taken into account in design and construction planning. On Notifiable Projects this sort of information will, as a matter of course, have been placed in the Health and Safety File but, where there is no File, such information should be located and made available to each duty holder including the Client.

However, not all 'As Built' drawings have to be incorporated. Only Health and Safety-related drawings go into (or are referred to within) the Health and Safety File. Non-Health and Safety 'As Built' drawings could go into the Building Manual or other handover documentation.

(e) Existing services: Accurate information on any existing services within or adjacent to the site should be identified, condition, routes and depths established, and any potential area of concern clearly identified within the Health and Safety information provided to a Designer, Contractor (on a Non-Notifiable Project) or Principal Contractor (on a Notifiable Project).

This must include, in particular, details on any existing 'live' services. Where demolition or significant dismantling works are planned, the isolation of such services must be established before any works are started.

If the locations of services are not known but are known to be somewhere in the site, investigations will be necessary to locate these.

(f) Ground conditions/site investigations/contamination reports: It is a requirement of the CDM Regulations that accurate information is made available to all Designers prior to the design of any project. This is not an 'aspirational' issue; it is mandatory, and a duty under the regulations that falls on the Client, in the first instance, and on CDM-Cs where they are available (on Notifiable Projects). Reports on ground conditions and contamination can provide invaluable information in relation to the elimination or control of potential risks during the works, and basic desktop studies will be essential starting points for many projects.

If the Client does not have the relevant information or has not already commissioned reports, early procurement of such reports will be of paramount importance in the information flow process and can be essential for Designers. This is especially the case when working on brown-field sites or where existing uses include storage facilities. These matters must be taken into account at the earliest stages of design if serious risks are to be eradicated or minimised.

(g) Asbestos Registers: Where works are being carried out within or adjacent to an existing structure, the Control of Asbestos at Work Regulations 2006 require that the owner of any structure has a written Asbestos (Management) Plan or similar document available for the property. This document should identify whether or not there are asbestos-containing materials within the property, where they are located and an assessment of their condition. This is particularly relevant where structures were constructed before the mid-1990s. If no such reliable document is available, it is essential that, in the case of demolition or major refurbishment works, a Type 3

For an existing structure within which demolition work or major refurbishments are to be carried out, ask the following:

- *Has an Asbestos Survey been completed? If yes…*
- *If it was a Type 2 Survey, was it complete enough to fully inform Designers and Contractors of the asbestos risks?*
- *Was it a Type 3 Asbestos Survey?*
- *If not, this should be completed prior to starting on site*

If no Asbestos Survey has been completed,

- *Arrange for a Type 3 Survey to be carried out (or possibly Type 2 depending on the extent of any refurbishment and planned disruption of existing materials)*
- *Make the results available to the CDM-C on a Notifiable Project and to the Designers and Contractors on a Non-Notifiable Project*
- *Make sure the Client has a copy of the results for inclusion in the structure's Asbestos Register*

In both cases the Asbestos Information, as amended to reflect any changes due to Notifiable or Non-Notifiable Projects, must be placed in an existing or any new Health and Safety File

Asbestos Survey is carried out before any operations are commenced by the Contractor within these parts of a structure. This documentation should be incorporated into the Project Information Resource and later amended to take account of any changes that result from construction work. This information will then be incorporated into the project, site or structure Health and Safety File. Again this is not an aspirational issue – it is mandatory.

Note: *Where operations are being carried out within domestic properties which are owned by Housing Associations or Co-operatives, or other landlords, there is an obligation to ascertain if asbestos is present and to provide the Asbestos Management Plan (Register) or, where appropriate, arrange for an appropriate survey. Such properties are not deemed to be 'domestic' for the purposes of the CDM and Asbestos regulations.*

6.6   When to Provide the Information?

The answer should now be obvious – information must be provided at any time and all times throughout the project stage from inception through to completion. Information has to be provided at appropriate times to keep the flow of relevant Health and Safety information moving so that 'the RIGHT INFORMATION goes to the RIGHT PEOPLE at the RIGHT TIMES'.

6.7   Information from Designers for Designers and Contractors

Designers are required to take all reasonable steps to provide, with their designs, sufficient information about aspects of the design of the structure and its construction and maintenance that are needed to enable others to comply with their duties, including those in relation to the Health and Safety File.

Designers therefore need to pass on information that results from DRM and provides residual risks (or suggested control measures) that other Designers need to know about, as well as the residual risks that need to be passed on to Contractors, and to do this so that the information is available in sufficient time to be taken account of by those who might need it.

In addition, project managers or lead Designers will normally issue copies of minutes or notes of Design Review Meetings, Project Team Meetings, Health and Safety Reviews, Risk Management Workshops, etc which should include any DRM decisions made as a result of such meetings.

These DRM outcomes will then be part of the Project Information Resource and will be used in various packages of information that have to be produced either by the CDM-C or a lead Designer (or Co-ordinator) on a Non-Notifiable Project.

6.8   Information for the Principal Contractor

When assessing information requirements for the Principal Contractor, it is necessary to ensure that the information provided is appropriate, meaningful and complete. Designers have to take all reasonable steps to provide sufficient information adequately to assist Contractors on all projects. On Notifiable Projects the CDM-C will liaise with the Principal Contractor to establish information that the latter may need to be able to prepare the Construction Phase Plan.

As examples of this, and in addition to the information identified above, thought must be given to:

- Identifying significant, unusual or difficult-to-manage residual risks that Contractors/ Principal Contractors will be required to manage and need to know about
- Requiring appropriate responses in tender documentation and appropriate tender assessments
- Highlighting the time allowed for mobilisation and mobilisation issues
- Requirements for delivering information to enable the Project Information Resource to be developed

6.9   Health and Safety File

It is crucial that Designers comply with their duties to assist in the compilation of the Health and Safety File so that it can be delivered when it is needed – at hand-over of the structure.

Designers do not need to provide risk assessments or details of residual risks relating to common risks that competent Contractors will be aware of, but do need to provide residual risks that relate to significant issues that Contractors and others will need to know about so that they can safely work on a structure in the future. Because of the Designer duty to provide information, and because there is now a regulatory requirement for Clients to keep Health and Safety Files up to date, this will mean that Designers must provide Clients with relevant residual risk information from Non-Notifiable Projects, just as the CDM-C will include relevant information in any new or existing Health and Safety File on a Notifiable Project.

This issue is discussed further in Section 8, which deals specifically with residual risk information and the Health and Safety File.

6.10   Ensuring the Flow of Information

There are many ways of ensuring that information flows successfully. These include – but are not limited to – the following:

Do...

- Meet face-to-face and discuss issues
- Work *with* other Designers and the CDM Co-ordinator who is the facilitator for co-ordinating Health and Safety issues on Notifiable Projects
- Make sure DRM issues are discussed in Design Review Meetings, Team Meetings, Health and Safety Co-ordination meetings and Workshops on Non-Notifiable as well as Notifiable Projects
- Contribute to the PIR and Hazard and Risk Register

Do not...

- Start design work until the CDM-C has been appointed on Notifiable Projects
- Work on your own without co-ordinating your work with others – especially on Non-Notifiable Projects
- Leave out, or forget, the CDM-C on Notifiable Projects
- Work on your designs without considering DRM *from day 1*
- Forget to carry out your DRM actions
- Provide risk assessments

6.11 ...When Information is Not Flowing Well...

- Don't stay quiet – no-one benefits
- Try to generate (improved) co-ordination between parties
- Seek agreement between parties to pass information to each other
- On Notifiable Projects, voice any concerns to the CDM-C
- On Non-Notifiable Projects, ask the lead Designer to seek improved co-ordination
- Keep a simple record of concerns
- As a last resort, voice concerns to the Client – especially on Non-Notifiable Projects

Risk Management in Design

7.1 Introduction: For All Construction Projects

Anyone who considers the way a structure is to be built is likely to be a Designer, and Designers must consider risk management for that design.

The degree to which this is carried out needs to be proportionate and appropriate to the complexity of risks associated with the activities required to build, maintain, clean, refurbish, dismantle or demolish the elements of the structure that is being designed.

Health and Safety risk management in design, called Design Risk Management (DRM) in this Guide, is the terminology used to describe the management systems, processes and procedures that provide for or allow a systematic approach to ensuring that all the appropriate Health and Safety issues are identified and dealt with for each structure comprising a construction project. DRM should not, however, be seen as separate from other design processes, but should be embedded into and integrated with normal design procedures.

DRM addresses the Health and Safety aspects of design and planning in project management during the design process. While much design is naturally in advance of site works, it usually continues through the construction phase because the design programme is continuing after commencement on site and/or because of changes that are taking place due to, for example, site conditions, value engineering, innovation, Client requirements, design team revisions or Contractor's design changes.

Virtually no project is achieved without revisions, amendments or changes to suit site conditions or other circumstances, and those responsible for these revisions and amendments are also carrying out design work. DRM, and all it entails, therefore continues until the design is physically translated into a completed structure.

7.2 Design Issues and Health and Safety

Designers automatically ask themselves a number of questions when they approach a design, and these can all have an impact on Health and Safety risks for those constructing or working on a structure after completion. These questions can include:

- What is being built, what should it look like and what is its function?
- Where is it being built, and what is adjacent to it?
- What materials should be used?
- How will it be constructed?
- When will it be built?
- How long will it take to build and when is it required?
- What other constraints and circumstances can affect the design?
- Who else is designing this project and what elements or aspects are they addressing?
- What will be continuing to be done or used on the site during the construction phase?

Since the introduction of the CDM Regulations, it has also been a legal requirement to address another series of questions which have a specific Health and Safety dimension (although, previously, some Designers may have been considering these questions anyway):

- What competence do we need for this design work, and what is the competence of those we are working with or who are working for us?
- How will this structure or element be maintained safely?
- How will this be cleaned, accessed, altered, refurbished, removed or demolished safely?
- What is the competence of those who will carry out the above works?
- What information will all these people need to do that work safely?

The CDM Regulations 2007 now also make it a legal requirement to consider the following questions:

- Is this a structure that will be used as a workplace?
- If it is, has it been designed to take account of the provisions of the Workplace (Health, Safety and Welfare) Regulations 1992 and the 2002 Amendments?

DRM ensures that all the questions outlined above (and others) that have any known or foreseeable effect on the Health and Safety of workers, those affected by the work, or users of a workplace are addressed appropriately and proportionately to the size and complexity of the works involved on the project.

7.3   Avoiding Foreseeable Risks

A common misconception among some Designers since the introduction of the CDM Regulations has been to think that the DRM process requires them simply to identify the residual hazards in their designs and to provide information so that others (usually the Contractor) can deal with the risk issues. In fact, it is essential that Designers direct their actions towards eliminating hazards and, if they cannot, to reducing risks through design decisions or provisions. Only as a last resort should Designers seek to rely on actions by the Contractors on site to provide the means of protection of workers and others during work activities.

Designers are required to avoid foreseeable risks 'so far as is reasonably practicable, taking account of other design considerations'. The greater the risk, the greater the effort that must be given to eliminating or reducing it. This does not mean that designs have to be 'zero risk' or that Designers need to consider risks that cannot be foreseen. What it means is that Designers have to produce designs that can be constructed, maintained, used or demolished in reasonable safety and without unreasonably affecting the health of those working in construction. (Demonstration of what is reasonable will rely on some sort of written audit trail by the Designer of what factors they considered at the time of their design – see Section 9.)

7.4   All Risks or Significant Risks?

In carrying out risk management, the emphasis should not just be on significant issues or significant risks. Designers are in a unique position to improve the whole range of construction-related Health and Safety issues and must not overlook the contribution they can make to reducing ill health and minor injuries, especially those that relate to such issues as hand-arm vibration, noise-induced hearing loss, musculo-skeletal disorders, and dermatitis.

Designers must make sure that they consider and address 'low-probability, high-risk' issues that could cause great harm or harm to a large number of people. These construction risks cannot be ignored even if the probability of occurrence is low; the possible consequences require them to be considered, eliminated if possible, but reduced and controlled if not.

Designers need to recognise that some hazards that may at first be thought trivial because they are considered as resulting in low risk could actually present significant risks as a result of changes to site or construction circumstances, or methods or programming choices that differ from the Designer's assumptions. Some 'low' risks might therefore need to be kept under review during design development in case circumstances change.

Whilst Designers must never fail to give appropriate efforts to eliminating or reducing risks relating to unusual aspects of a project and to factors generally recognised as significant, they should consciously avoid producing information for others about hazards that will not involve significant risks, or hazards of which a reasonably competent Designer or Contractor can be expected to be aware. If there are a few – or even no – significant residual issues that anyone needs to know about, the Designer should state this. Producing excessive or unnecessary information just because of a fear of appearing to have not done a good job or 'covering one's back', is counterproductive and may obscure the really important information.

Significant hazards and residual risks on one or two sides of A4 are likely to be far more 'life saving' than 25 or 30 pages of 'risk assessments' – and they can be far more cost effectively produced

7.5   Starting Design Work

Where a project is Notifiable, Designers are not permitted to start design work (other than initial design work) in relation to any project unless a CDM-C has been appointed. This ensures that the process of providing the necessary risk information and co-ordination that is essential to effective DRM is put in place before design work begins. To comply with this requirement, Clients and Designers need to know how initial design work is defined and, in this regard, neither the Regulations nor the ACoP provide a finite definition (ACoP paragraph 66 is the most relevant and is cited in Section 2 above).

Definition of initial design work

Initial design is considered by the APS to be 'enough work to allow the Client to appraise their business and project needs and objectives to enable the Client or their advisors to decide whether or not to proceed with a construction project'. This takes account of the understanding that the business and project needs and objectives would be those provided or described in a strategic brief that addresses such issues as function, mix of uses, scale, location, quality, cost, time, safety, health, environment and sustainability of a proposed project. Business and project needs and objectives could include:

- Feasibility study
- Development of a business case
- Reviews of possible constraints on development
- Sufficient exploration of project implications (design) to enable the Client or Designer to know whether or not the probable construction phase is likely to make the project Notifiable or Non-Notifiable

The APS view is made in the context of the Regulations and ACoP as a whole and based on the fundamental reasons for early appointment of the CDM-C. Initial design work *should not be inferred* to include:

- Initial designs for the project
- Competition designs
- Schematic proposals for a site that establish organisation, structure, construction and appearance of structures and accommodation so that later changes that might result from risk-related considerations are precluded or made difficult to incorporate
- More detailed designs

Initial design work should only relate to the very earliest explorations that indicate that a project has the potential to match an initial Client brief and notions of cost. Put simply, the project moves beyond initial design work when the design moves into concept development, concept design and the implementation of any strategic brief.

Appointment of the CDM-C at any later stage would frustrate the purpose of this appointment and also effective information flow, co-ordination and competence requirements, and place the Client in the position of CDM-C for the early critical phase of DRM. Commencement of even concept design work before appointment of the CDM-C would prevent consideration of the issues that the CDM-C has to deal with and prevent the effective discharge of Client responsibilities.

This means that, where the construction phase is projected to be over 30 days or 500 normal person days of construction work, the following examples of design work will require the appointment of a CDM-C:

- Preliminary design to inform Public Consultation or Parliamentary Approval or Consent
- Concept, preliminary, or outline design, even if money for construction is not yet known to be available or if permissions have not yet been applied for or received
- A design competition for a bridge, building, or other structure. [In this case the Client (instigator of the competition) needs a CDM-C to provide all the right information to each competitor, and each design competitor must comply with Designer duties. The CDM-C and Client will need to judge the entry, taking into account the Health and Safety information submitted by each design team.]
- Early or outline design to provide enough information for an Early Constructor Involvement (ECI), Private Finance Initiative (PFI), Design Build Finance and Operate (DBFO), Design Build Finance Operate and Maintain (DBFOM) or similar project to be put out to expressions of interest or tender
- Designs produced before the project is given to a Special Purpose Vehicle (SPV) to take forward
- Structural calculations
- Production of a specification
- Producing Bills of Quantities (the production of a schedule of rates may be design work depending on the instructions or constraints within the text)
- Planning for a demolition project
- Planning by a Contractor or project management organisation for any project where a 'Designer' (in the traditional use of the term) is not engaged

There may also be times when a Client's Code of Construction Practice or similar document strays into instructions or constraints that could be considered design.

> *Designers must always remember that, whatever is meant by 'initial design', the work that the Designer carries out in every phase is subject to Regulation 11 'Duties on Designers'*

7.6 Programmes, Timing and Co-ordination

A Designer, and particularly a lead Designer, will need to monitor progress in relation to the project or design programme and the needs of others who need the information and designs on which they are working. It is essential to:

- Determine design and (contractual) construction programmes, as well as scheduled completion arrangements and dates (sectional completion dates or early hand-over requirements must be taken into account)
- Identify the Designer's intended programme for design reviews and Health and Safety monitoring procedures
- Establish who will undertake the reviews and which design team members will be involved
- Synchronise monitoring tasks with scheduled design reviews, design stages, or in relation to design team meetings

It is often the case that circumstances change, so monitoring must be ongoing. Designers must review procedures, programme, and information continuously throughout the design development process – remembering that this will usually continue throughout the construction programme too.

Co-ordination and co-operation: It is essential that the various Designers on a project use compatible methods of information transfer, agree DRM protocols, and consider how their designs may interact in terms of Health and Safety. These issues (covered in Section 5) are critical to the effectiveness of DRM.

7.7 Tracking Design Changes

For each Designer, a means of tracking design changes and the resultant Health and Safety DRM changes will be essential on all but the very simplest and non-complex projects. It may also be necessary to be able to identify not only that changes have been made, but that risk appraisal and management have also been carried out and the resultant information given to those who need it.

In tracking design changes, the simplest method is likely to be most straightforward. A table or spreadsheet to log the 'date-currency' against each design team member and his/her residual risks can be useful, and use of unique reference numbers for risk items will be helpful in identifying the final status of any design issue and any residual risk information that will need to be passed on.

This tracking process may well be consolidated by the CDM-C on a Notifiable Project by the use of the project Hazards and Risk Register, but on Non-Notifiable Projects, unless the Client appoints someone, for example the lead Designer, to ensure co-ordination, each Designer will need to ensure that design change tracking takes place when it is necessary. The tracking system needs to be in place when design begins – especially where project hazards are likely to be multiple and complex.

This tracking system will also enable an effective audit trail (*see Section 9*) to be maintained. This will not only be useful in checking how and why earlier decisions were made when design change is necessary, but will provide an invaluable record and resource for others in the event that the design or project is 'shelved' or delayed for a number of months or years, whether or not the design or project passes to a different project team or Designer. The tracking system will also enable Designers to defend themselves in the event of an investigation.

7.8　The Design Risk Management (DRM) Process

Principal targets for DRM: Designers should be aiming to:

- Eliminate or minimise risks from site hazards
- Design out or minimise risks from health hazards
- Design out or minimise risks from safety hazards
- Design in features to reduce risks, for example, from working at height
- Simplify safe construction, maintenance and cleaning work
- Consider pre-fabrication to minimise high-risk work
- Ensure the suitability and compatibility of separate but interacting or inter-relating designs
- Take into account the Workplace (Health Safety and Welfare) Regulations (and Amendments)
- Provide information on significant risks associated with their design at pre-arranged stages during the project design to those who need it

- Provide relevant supporting information (for example, by ensuring that copies of minutes of all design team meetings are circulated expeditiously)
- Identify any hazards applicable to future work, including cleaning, maintenance and demolition for the Health and Safety File

Integrate with overall design process: One essential aspect of DRM is that these targets should be achieved by processes that are an integral part of the overall design process, and not 'bolt-on' processes that are delayed or deferred until the design is (virtually) finalised.

Management procedures: The starting point is the setting up of appropriate management procedures. The more complex the project, the more systematic these procedures need to be.

On Notifiable Projects the CDM-C has the duty to ensure that co-ordination arrangements are made, that information is made available to those who need it, and that Designers comply with their duties. Establishment of management procedures will therefore be assisted by the involvement of the CDM-C. On the other hand, on Non-Notifiable Projects, as has been previously pointed out, the Client has to ensure that co-ordination arrangements are made and that information is made available to those who need it; the Designer may be expected to assist with this.

Management procedures which a Designer may assist a Client to establish include:

- Co-ordination of work with others involved using perhaps a lead Designer to carry out co-ordination and ensure information flow on Non-Notifiable Projects or by co-operative working with the CDM-C, Client, Contractors and other Designers on Notifiable Projects
- Ensuring DRM throughout (design) feasibility, concept design, detailed design and design completion stages
- Hazards and risks being identified, designs amended and appropriate risk management procedures put in place
- Proportional responses to foreseeable significant risks
- Adequate time allowances for CDM procedures

- Systematic design reviews at suitable stages checking for hazards, hazardous activities and Health and Safety management responses
- Checking that elements of designs are both suitable and compatible
- Adequate records of the risk management process
- Hazard and risk information being checked before design information is issued
- Hazard and risk information being provided in a clear, concise, and appropriate format
- Information being passed on to those who need it, when they need it – *even though they may not know they need it* – including to other Designers, those tendering, and to Contractors (and to the CDM-C when the project is Notifiable)
- Procedures to deal with other project-specific or unique issues

Construction knowledge: For DRM processes to be effective, Designers need to have a clear understanding of how structures and their elements are to be constructed, cleaned and maintained, and they must be able to identify at least one safe and healthy way for their design to be constructed. These issues require the Designer (or appropriate members of the design team) to be experienced in these areas, preferably by having been on site while similar works were underway.

At least one member of each design team needs to be applying some valid experience of construction and design to the DRM process.

Location and surroundings: For Designers to be able to carry out DRM effectively, it is essential that they have an understanding of the site, the proposed construction, the local area and circumstances (including operations and sites in the locality) so that they are able to identify the apparent and inherent hazards and related risks that arise as a result of possible interactions between these. This will normally require a visit to the site to check on inherent site-related hazards and risk issues to ensure that they are familiar with the context of the project.

Particular attention must be given to adjoining construction sites, and early co-ordination and co-operation with those involved will be necessary so that risks that may occur as a result of interaction between these operations can be eliminated, minimised and controlled. On projects with longer programme times, Designers – or perhaps the CDM-C on Notifiable Projects – must monitor what is going on around and

Designers need to understand how the structure can be constructed, cleaned and maintained safely. This involves:

- Taking full account of the risks that can arise during construction, giving particular attention to new or unfamiliar processes, and to those that may place large numbers of people at risk
- Considering the stability of partially erected structures and, where necessary, providing information to show how temporary stability could be achieved during construction
- Considering the effect of proposed work on the integrity of existing structures, particularly during refurbishment
- Ensuring that the overall design takes full account of any temporary works, for example, false work which may be needed, no matter who is to develop those works
- Ensuring that there are suitable arrangements (for example, access and hard standing) for cranes and other heavy equipment, if required

about their projects and ensure that they seek the co-operation of those responsible for any proposed construction works due to commence during the life of their project. This may be a resource-demanding issue on projects that cover several miles – for example, trams, highways, pipelines, sewers and drainage schemes.

Process and information: On any project the application of Health and Safety risk management to designs will require:

- Identification of other Designers
- Listing of the structures or elements of structures
- Identifying the construction activities necessary (or likely) to be used in building, cleaning, maintaining, altering and dismantling or demolishing those structures or elements
- Identifying adjacent construction activities or sites
- Identifying critical neighbouring or off-site issues arising from hazardous activities, locations or 'at risk' populations
- Identifying the hazards or hazardous activities

- Amending the design to eliminate those hazards or hazardous activities if feasible
- Amending the design so that risks are reduced from the remaining hazards
- Identifying ownership of issues to ensure actions are taken
- Ensuring issues are 'closed-out' before information is finalised and issued
- Providing the relevant information to those who need it – the CDM-C on Notifiable Projects and the Client, other Designers and Contractors on all projects
- Recording the essentials of these processes so that hazards and risks can be tracked

The effort given to the above needs to be appropriate and proportionate to the nature of the design and to the nature and complexity of the Health and Safety issues and, of course, it almost goes without saying that these processes will generate associated information that will be essential pre-construction information and a significant part of the Project Information Resource.

(a) Identifying other Designers: This can be achieved by identifying the various structures or elements of design and then listing who is designing each. Another way is by discussing who is doing what with each team leader and party involved. Having identified who is working on each aspect of design, contact person or persons within each design team need to be identified and a programme and possible meeting dates established to discuss Health and Safety issues and the sharing of information.

(b) Listing the structures or elements of structures that each Designer is responsible for: This will help ensure that all aspects of the project are identified and that someone is allocated responsibility for each 'deliverable'.

(c) Identifying hazards, hazardous activities and associated risks: For each structure or element, it will be essential to identify foreseeable hazards associated with the activities necessary to build or erect it, to clean and maintain it, to alter or add to it, and to eventually dismantle or demolish it. This will enable a Designer to then consider how designs or structures can be changed or amended to avoid or reduce the occurrence of those activities.

Table 7.1 provides a list of 14 basic hazards that can occur on construction sites and can be used as a quick check when identifying construction hazards on projects – but

Table 7.1: 14 basic hazards in construction work

Working:

- At height
- Within or adjacent to moving vehicles or objects
- Where there is the danger of collapse or instability
- Over or adjacent to water or other fluids
- With or close to electricity
- With or adjacent to flammable materials with danger of fire or explosion
- With or close to high energy or pressure or temperature sources such as pipes or vessels
- With materials or substances that are toxic or otherwise hazardous to health, including dust and those materials or substances that release hazardous vapours or fumes or which deplete oxygen
- With heavy or awkward loads
- With lifting devices
- In restricted space or positions or requiring repetitive actions that strain the muscles, causing musculoskeletal disorders
- In a noisy, hot or cold environment
- With powered tools or other equipment, causing Hand Arm Vibration (HAV)
- Injuries in confined spaces

it is pointless for Designers simply to list the hazards that occur on their project, for generic information is worthless. What must be provided is *task or process specific information* so that the relevant aspects of construction can be considered and associated risks eliminated, reduced or managed.

One practical way to achieve this is for a Designer to consider each work activity necessary to build, maintain, clean, or in any way work on the structure or element, and assess if any of the 14 hazards will apply. If the hazard is foreseeable, then the Designer must ask what can be done to eliminate the hazard or reduce the risks arising from it?

(d) Eliminating hazards: For each hazard or hazardous activity, Designers need to consider if there are ways of eliminating that hazard by making alternative design decisions. Can the design itself be altered? Can different materials be specified? Can the form of construction be changed? There are many ways in which Designers can make changes, and some effective examples are listed in Table 7.2.

Table 7.2: Hazards to consider in design

A Designer should where possible:

1. **Select the position and design of structures to minimise risks from site hazards**, including:

   - Buried services, including gas pipelines
   - Overhead cables
   - Traffic movements to, from and around the site
   - Contaminated ground, for example minimising disturbance by using shallow excavations and driven, rather than bored, piles

2. **Design out health hazards**, for example by:

   - Specifying less hazardous materials, e.g. solvent-free or low solvent adhesives and water-based paints
   - Avoiding processes that create hazardous fumes, vapours, dust, noise or vibration, including disturbance of existing asbestos, cutting chases in brickwork and concrete, breaking down cast in-situ piles to level, scabbling concrete, hand digging tunnels, flame cutting or sanding areas coated with lead paint or cadmium
   - Specifying materials that are easy to handle, e.g. lighter weight building blocks
   - Designing block paved areas to enable mechanical handling and laying of blocks

3. **Design out safety hazards**, for example:

   - The need for work at height, particularly where it would involve work from ladders, or safe place of work is not provided
   - Fragile roofing materials
   - Deep or long excavations in public areas or on highways

   Continued

Table 7.2: *Continued*

- Materials that could create a significant fire risk during construction
- Reducing depths of foundations or drainage, where feasible

4. **Consider prefabrication** to minimise hazardous work or to allow it to be carried out in more controlled conditions off-site, including, for example:

  - Designing elements, such as structural steel work and process plant, so that sub-assemblies can be erected at ground level and then safely lifted into place
  - Arranging for cutting to size to be done off-site, under controlled conditions, to reduce the amount of dust released and noise produced

5. **Design in features that reduce the risk of falling/injury** where it is not possible to avoid work at height, for example:

  - Early installation of permanent access such as stairs to reduce the use of ladders
  - Edge protection or other features that increase the safety of access and construction

6. **Design to simplify safe construction**, for example:

  - Providing lifting points and mark the weight and centre of gravity of heavy or awkward items requiring slinging, both on drawings and on the items themselves
  - Making allowance for temporary works required during construction
  - Designing joints in vertical structural steel members so that bolting up can easily be done by someone standing on a permanent floor, and by use of seating angles to provide support while the bolts are put in place
  - Designing connections to minimise the risk of incorrect assembly
  - Designing structural alterations so that existing structural support can remain in place until new support has been installed

7. **Design to simplify future maintenance and cleaning work**, for example:

  - Making provision for safe permanent access
  - Specifying windows that can be cleaned from the inside

Continued

Table 7.2: *Continued*

- Designing plant rooms to allow safe access to plant and for its removal and replacement
- Designing safe access for roof-mounted plant and roof maintenance
- Making provision for safe temporary access to allow for painting and maintenance of facades, etc. This might involve allowing for access by mobile elevating work platforms or for erection of scaffolding

8. **Identify future demolition hazards for inclusion in the Health and Safety File**, for example:

- Sources of substantial stored energy, including pre- or post-tensioned members
- Unusual stability concepts
- Alterations that have changed the structure

Note: This is not an exhaustive list, nor is each item relevant to every project

(e) Reducing risks: Where hazards or hazardous activities cannot be eliminated, Designers must consider what design decisions can be taken that reduce the risk(s) arising from these hazards or activities. This is a crucial aspect of design, and it can be achieved by any, or all, of the following:

(i)   Altering the design to reduce the likelihood of risk
(ii) Changing the design to reduce the severity of risk
(iii) Changing the methods, programme or sequence of work to reduce risk
(vi) Reducing the numbers of people exposed to the risk
(v)   Changing the frequency or need for a risky activity, especially in cleaning and maintenance

It is clearly essential for the Designer to have a good working knowledge and appreciation of the Health and Safety regulations that apply to construction and construction activities so that hazards, hazardous activities, and risks cannot just be identified but be dealt with appropriately.

Designers CAN make a difference

(f) Identifying ownership: To ensure that risks are controlled, it is necessary that each hazard or risk has an identified 'owner', who will follow through on the options and possibilities for avoidance or risk reduction and ensure that risks are dealt with (so far as is reasonably practicable), taking into account other design considerations. This 'ownership' will also enable effective co-ordination with others affected by the decisions being made who, for example, could be: other Designers, the Client, the Principal Contractor, Contractors, as well as those managing adjoining sites or properties, maintenance teams or facilities managers.

(g) Significant residual risks or issues: In following the processes listed above, many Health and Safety issues will be dealt with, some hazards eliminated, and many risks reduced. The next stage of the process is to identify which remaining issues (hazards and hazardous activities) are significant.

These significant risks will be those that a competent Designer or Contractor cannot be expected to know about as well as those that may be unusual or difficult to manage.

Information about significant hazards and related significant risks that remain in a design, together with any assumptions made by the Designer about working methods or precautions, must be provided to the right people at the right times and in an appropriate form (*see Sections 6 and 8*).

Those preparing this information need to make sure that only significant risk issues or 'difficult-to-manage' elements of construction are referred to, so that the important issues are not obscured by less critical information. Only project-specific Health and Safety information is required, and this should be placed where it will be of most benefit to future users.

Examples of issues that are always considered to be significant and on which Designers would therefore always need to provide information (if relevant to their design and their project) are given in Table 7.3.

It is an essential aspect of DRM that information on significant hazards and risks that remain in a design are identified by Designers and that this information is passed on to

Table 7.3: Examples of significant hazards where Designers always need to provide appropriate information

- Hazards that could cause multiple fatalities to the public, such as tunnelling or the use of a crane to a busy public place, major road or railway
- Temporary works required to ensure stability during construction, alteration or demolition of the whole or any part of the structure, e.g. bracing during construction of steel or concrete frame buildings
- Hazardous or flammable substances specified in the design, e.g. epoxy grouts, fungicidal paints or those containing isocyanates
- Features of the design and sequences of assembly or disassembly that are crucial to safe working
- Specific problems and possible solutions, e.g. arrangements to enable the removal of a large item of plant from the basement of a building or means of access to, and safe maintenance of, items in confined spaces
- Structures that create particular access problems such as domed glass structures
- Heavy or awkward prefabricated elements likely to create risks in handling
- Areas needing access for which a safe solution is not obvious, e.g. where normal methods of tying scaffolding may not be feasible, such as façades that have no opening windows and cannot be drilled; high level glazing

Note: This is not an exhaustive list, nor is each item relevant to every project

those who need it. The whole process is worthless if this aspect is not successfully achieved in time for the right people to use the information.

(h) Monitoring issues to ensure designs are not issued prematurely: One of the consequences of changing designs – to eliminate hazards or to reduce risk – is that changes to reduce one risk can result in increased risk elsewhere. It is therefore important to have in place a method for ensuring that hazards that have not been 'signed off' are periodically reviewed so that design information can be updated, the balance between risks reconsidered and then appropriate residual risk information issued. To do otherwise could result in premature issue of information.

(i) Issuing residual risk information: On a Notifiable Project, each Designer has to pass information on significant residual risks to the CDM-C, who then ensures that Health and Safety information is made available to those who need it and places appropriate information in the Project Information Resource and, where appropriate, in the project Health and Safety File. This process can be expedited by the direct transfer of information between Designers, provided always that the CDM-C knows what is going to whom and can take account of its possible implications for others.

On Non-Notifiable Projects the residual risk information will need to be passed to those who need it – other Designers, Contractors and the Client – by the Designers generating it or perhaps to another person appointed by the Client to co-ordinate the project.

Closure: Someone must ensure that hazards that have been dealt with are 'signed off' by the Designer or 'owner' and that there are procedures to record what information has been passed on to whom and when. This does not mean that every Health and Safety issue has been eliminated or a solution found. Where residual risks remain, 'closure' will mean ensuring that the relevant information has been passed on to those who need it. On a Notifiable Project the arrangements for this process will be overseen by the CDM-C, while on Non-Notifiable Projects someone else – the Client, the lead Designer or whoever is assisting the Client – will have to check that it happens.

7.9 Taking into Account the Workplace (Health, Safety and Welfare) Regulations

Under CDM 2007, Designers have to consider Health and Safety issues relating to the use of *any* structure they design if it is to become a workplace. Designers responding to this new CDM duty (Regulation 11(5)) have to take account of the provisions of the Workplace (Health, Safety and Welfare) Regulations 1992 (the 'Workplace Regulations'), last amended in 2002, when designing a structure or the materials used in it. Many structures will have to comply with Building Regulation or Warrant requirements, but many would not be covered by such legislation. Designers therefore need to be aware that this new CDM requirement relates to the use of the designed structure as a workplace and does not directly relate to building standards' requirements.

In summary, the Workplace Regulations apply to any non-domestic premises made available to any person as a place of work and include any room, lobby, corridor, staircase, road (but not including a public road) or other means of access to or egress from any such premises.

Note:

1 'Premises' has a wider definition than simply a structure or a building, for example the car park at a factory would form part of that workplace so 'workplace' can logically include certain outdoor areas also, and

2 Workplace and premises apply to 'temporary' as well as 'permanent' work sites.

The Workplace Regulations have their origins in the earliest welfare-based legislation in the UK dating from the 19th century and include important basic provisions. Some aspects are more specific, and a Designer or design organisation will need to have access to the detailed requirements of the Regulations and the related Approved Code of Practice (ISBN 0717604136).

In summary, the Welfare Regulations establish duties with respect to the following (the number references correlate with the Regulations):

4A Stability and Solidity: If a workplace is in a building, the building has to be appropriately stable and solid for the nature of use of the workplace.

5 Maintenance of the workplace and of equipment, devices and systems: A workplace and its equipment and systems must be properly cleaned and maintained and need to be designed to be *capable* of being maintained and cleaned at suitable frequencies. This includes emergency lighting, fixed window cleaning equipment, escalators etc.

6 Ventilation: An enclosed workplace must be adequately ventilated with fresh (uncontaminated) air without excessive drafts, smells or humidity and without risk of Legionnaire's disease.

7 Temperature in indoor workplaces, thermal insulation and avoidance of effects of sunlight: Workplace temperature must be reasonable and the workplace adequately thermally insulated where necessary (with regard to type of work and physical activity or persons). Excessive effects of sunlight on temperature are to be avoided.

8 Lighting: Workplaces must be adequately lit, and so far as is reasonably practicable this should be by natural lighting with suitable emergency lighting where loss of lighting would present a risk.

9 Cleanliness and waste materials: Surfaces within a workplace must be capable of being kept sufficiently clean.

10 Room dimensions and space: People need sufficient room in relation to the number of people working there whilst they work.

11 Work stations and seating: People need to be able to work in adequate comfort (including protection from the weather) and able to work without strain.

12 Construction of floors, handrails and traffic routes: Workplace floors and surfaces of traffic routes (walking or vehicles) need to be safe and fit for purpose (consideration of uneven/slippery surfaces, holes etc).

13 Preventing falls and falling objects: Mainly revoked by Work at Height Regulations 2005, but securely cover/fence off any tank/pit/structure where a person may fall into a dangerous substance.

14 Materials used for windows and transparent or translucent doors, gates and walls, shall be of safety material or protected from breakage where necessary and marked to make them visible.

15 Safe opening and position (when open) of windows, skylights and ventilators.

16 Ability to clean windows etc. safely.

17 Organisation etc. of traffic routes: People and vehicles need to be able to circulate safely. This applies to traffic routes including stairs, ladders, ramps etc. and loading bays together with associated doors and gates and includes suitable measures to separate vehicles and pedestrians.

18 Doors and gates able to be safely opened with safety features where needed and with a view of the space on the other side.

19 Escalators and moving walkways and their safety features: They must function safely with suitable safety devices and emergency stop controls.

20 Suitable, sufficient and readily accessible sanitary conveniences with adequate ventilation and lighting of the rooms containing them.

21 Suitable and sufficient washing facilities: In addition to being provided at toilets and changing rooms, suitable and sufficient readily accessible washing facilities (including showers if necessary) with hot and cold, or warm water should be provided and the rooms containing them provided with adequate ventilation and lighting.

22 Supply of drinking water: People must be provided with 'wholesome drinking water' which is readily accessible at suitable places and marked as fit for drinking.

23 Accommodation for clothing, including the drying of clothing where necessary.

24 Facilities for changing clothing and provision of seating: Provision of changing facilities if people need to change into special clothing for work with separate facilities (or separate use of facilities) for men and women if necessary. Facilities to be easily accessible, of sufficient capacity, and provided with seating.

25 Suitable, sufficient and readily accessible facilities for rest, for eating meals (if food could be contaminated in the workplace); facilities for expectant and nursing mothers; rest rooms or areas with adequate number of tables and seats with backs and provision for disabled workers. Arrangements need to include protection of non-smokers from tobacco smoke (except Scotland where smoking in the workplace is illegal).

25A Parts of the workplace organised to take account of disabled persons, especially doors, passageways, stairs, showers, washbasins, lavatories and work stations. Those parts of a workplace used/occupied by disabled persons at work need to take account of such persons (e.g. provision of doors, passageways, stairs, showers, washbasins, lavatories, workstations etc.).

There are real opportunities for Designers, from the early design phase right through to detailed design and specification, to make a difference with respect to these regulations.

Note also that, when looking at these Regulations, they are written to place duties on the relevant employer or person in control of a workplace to respond to the essential basic provisions highlighted above. Under CDM 2007, Designers have to respond to these 'employer-based' duties as they design and they need to be aware that certain solutions could compromise the position of the Client procuring the design, or other users or occupiers of the workplace, with regard to compliance with the Workplace Regulations.

7.10 Health and Safety on Construction Sites

Designers also need to take account of the issues raised by Part 4 of the CDM Regulations which relate to the duty of those who 'control' the way that construction work is carried out. Whilst the 'final' responsibility for compliance with Regulations 26 to 44 will fall to Contractors, on occasions Clients and Designers may have to decide what has to be done and how, and will have to take account of these Regulations.

However, there is a far greater responsibility on those who design and plan for work. Designers, and this can be Clients, Contractors and PCs too, can have significant influence, during the design process, on matters covered in Part 4. It is the Designer who, for example, decides on the depth and location of a drainage run, and therefore dictates the circumstances for its excavation, and these design decisions may leave the Contractor with significantly different problems in complying with Regulation 31 – excavations. The Designer or Client may decide on the layout and size of the footprint of the structure in relation to the site and its surroundings, hence significantly affecting possible traffic routes and access for emergency services, leaving the Contractor with possible difficulties in complying with Regulations 36 and 40.

Designers therefore have significant opportunities to make a difference in their designs, particularly with regard to the following Regulations in Part 4:

26 Safe places of work
28 Stability of structures*
31 Excavations
32 Cofferdams and caissons
34 Energy distribution installations
35 Prevention of drowning
36 Traffic routes*
37 Vehicles (insofar as enabling a suitable layout)
38 Prevention of risk from fire etc.
40 Emergency routes and exits
42 Fresh air*
43 Temperature and weather protection*
44 Lighting*

(*these require cross reference to the Workplace (Health, Safety and Welfare) Regulations)

7.11 Passing Information to Others

Getting essential information to those who need it: Designers use specifications, bills of quantities and (especially) drawings to convey residual hazard and risk information as well as any specific methods or sequences of construction they have identified. Much of this will need to be part of the information flow and information packages that will be used by others, and this type of information will be essential to the Principal Contractor when preparing the construction phase plan. Some residual hazard and risk information, including any specific methods or sequences of removal or demolition, will be needed for preparing or updating the Health and Safety File.

Often the best way of getting the right information to the right people – those working on construction – is to show significant or unusual hazards on design, production and record drawings using hazard symbols and a 'SHE' (Safety, Health and Environment) panel to convey hazard information. This should be placed in a common location with a standard format that includes space for critical Health and Safety information and related locations, elements or components.

Annotations should, of course, be cross-referenced to any applicable hazard and risk registers, information packs and the Health and Safety File. Any other risk control information that needs to be considered by others should also be referenced.

Avoiding unnecessary information: It is unnecessary for a Designer to carry out and produce design risk assessments of the residual risks associated with their design. What is needed is information about the significant residual risk and any assumptions the Designer may have made for the control of that risk (for example, a specific method or sequence of construction).

Supporting information: This will often be useful to enable others considering design change at a later stage to understand what the Designer considered the essential risk issues to be. Supporting information could also include hazard and risk management registers showing residual risks, or even the process by which some design decisions were reached. These may assist those on site so that they do not inadvertently change a design back to an earlier choice, thereby reintroducing the associated hazards that had previously been designed out.

Supporting information may also include notes of the identification and DRM process, meeting notes, or checklists and records that indicate how information has been passed on, when and to whom. On a Notifiable Project, each Designer needs to discuss the types and format of supporting information with the CDM-C who has to identify and provide information to those who need it.

7.12   Monitoring and Reviewing the DRM process

Throughout the design process, which can last a long time on some projects, Designers will need to review the application of the DRM processes to make sure that their 'eyes are not off the ball'. Some simple questions will help with this:

- Is enough being done during design to eliminate hazards and reduce risk?
- Is adequate and proportionate information about residual risks being generated and passed to the right people at the right times?
- Are any obvious or foreseeable issues being omitted?
- Are there issues that we are dealing with that other Designers need to know about and take into account?
- Are we wasting time deciding between options that have similar levels of risk rather than finding options that can reduce risk significantly?
- Are there elements and structures and issues being designed by others that we need to know about because we can foresee that there is a Health and Safety interface that we need to be dealing with?
- In the process of marking-up drawings with information or passing on information, are we correctly focusing on significant risks and 'difficult-to-manage' processes rather than generic risk issues?
- Are we worrying too much about insignificant hazards or those a Contractor manages every day without difficulty (for example, pointing out that working on a tall structure or a bridge has the hazard of working at height; or pointing out standard control measures such as wearing hard hats or high visibility jackets; or pointing out that works on a structure in a river involves the risk of working near water; or pointing out the hazard of moving traffic to a highways contractor)? The Designer should be concentrating on avoiding the hazard, or reducing the risk in their design, not on pointing out the obvious to other competent dutyholders.
- At all times, Designers need to check that the information is being reviewed and allocated appropriately for inclusion with pre-construction information (such as information packs) and/or the Health and Safety File.

Beware checklists: Checklists and pro forma systems should, at best, only be used as 'aids' to risk identification and management, not as Health and Safety information in their own right. At worst, the use of checklists, pro formae and tick boxes leads to the DRM process being done automatically and superficially, possibly leading those who use them to forget to consider site-specific issues.

Designers also need to avoid using generalised DRM procedures such as:

- Risk assessments based on standard industry-wide statements of hazard and risk
- 'Imported' checklist risk assessments
- Risk assessments and associated paperwork as a substitute for risk management

This last point needs to be kept uppermost in the Designer's mind. The issue is risk elimination, reduction and control – NOT risk assessment.

> *Risk elimination, reduction and control are the raison d'etre of DRM – not risk assessments!*

7.13 The Bottom Line

Throughout the whole DRM process, and particularly in deciding on the information to provide, Designers need to focus on:

- Using lateral thinking to identify the 'significant' or 'difficult-to-manage' hazards and associated risks on a project
- Arriving at informed judgements
- Producing useful Health and Safety information
- Making sure this information gets to the people who need it

For this DRM process to be effective, it is essential that each Designer's role, responsibilities, actions and interfaces with others are clear and well understood by all involved. Co-operation and co-ordination will also be essential aspects of achieving a successful end result – the minimisation of risk to those working in construction, by design!

Residual Risks, the Health and Safety File and Client Requirements

8.1 Residual Risks and Health and Safety Files

Availability of the Health and Safety File: The Health and Safety File for a structure must be available so that any future work can be designed, planned and carried out with all the significant health and safety issues known in advance. It is crucial therefore for Designers to comply with their duties to provide information with their designs and, on Notifiable Projects, to assist in the compilation of the information for the Health and Safety File so that it can be prepared and delivered when it is needed – at hand-over of the structure.

Residual risks: Residual risks are those risks that have been identified during design and construction work, have not been eliminated and which will remain in the structure or pertain to it after completion of the current project. The CDM Regulations require that information that relates to these residual risks is placed, together with any other relevant Health and Safety information, in the Health and Safety File. These residual risks can be embedded or intrinsic to the project and may also relate to the site, adjacent sites and surroundings, and their potential hazards and risks. Designers therefore need to recognise that the aim of the Health and Safety File is that it should contain all the appropriate information needed to allow future construction work and maintenance to be carried out with the health and safety issues identified and properly managed and for workplaces (as defined under the Workplace (Health, Safety and Welfare) Regulations 1992 and as amended by SI 2002/2174)) to be healthy and safe. This will ensure that they consider widely enough the sort of information that they need to pass on.

Information that Designers need to pass on for the Health and Safety File will derive mainly from their own DRM work – the residual risks that they know will remain or be the consequence of particular design decisions. For instance, pre-tensioned structural elements that will be hidden within a structure, or perhaps high-level window cleaning methods, high-level maintenance and access risks are all residual risks or information that need to be included in a Health and Safety File. On the other hand, there will be information on known hazards in or adjacent to an existing structure or site and some from hazards that Designers have decided should not be removed – for example, asbestos that is safer being left and covered up rather than removed from an occupied building, or contaminated land deliberately left in place – that will also need to be included.

Health and Safety risk information for the Health and Safety File can therefore come from a wide range of sources that may include:

- Existing Health and Safety File(s)
- Existing or new surveys (*see Section 6*)
- The Client
- Existing Hazard and Risk Registers
- 'As Built' drawings of existing structures (that contain relevant Health and Safety information)
- Residual risk information from the project designs
- Temporary works left in place – particularly if these can be used later for future work on, or removal of, the structure or elements of it
- Site-related hazards discovered before or during construction work
- Information relating to the ground-materials exposed or discovered pipes or cables, etc as found by the contractors
- Problems encountered during construction (and it may also be relevant to include information about how these were dealt with, particularly if materials or temporary works were left in place)

It is important to remember, however, that the Health and Safety File is not a Building Manual nor a Maintenance or Operations Manual and should only incorporate information relative to Health and Safety *regardless* of what any Client may request. If Health and Safety information is to be related to any handover documentation, for example a Building Manual, the Health and Safety File component must be easily identified and not be obscured by a forest of information.

As a simple guide, information should be:

- Project-specific
- Appropriate
- Cover information, including any residual risk information, that may be needed by those who may carry out maintenance, cleaning, demolition work or the design of further construction work related to the structure

Risk assessments: Designers do not need to provide risk assessments or details of residual risks relating to common risks that competent Contractors will be aware of, but do need to provide information about residual risks that relate to significant issues that Contractors and others need to know about so that they can work safely and without risks to health on a structure in future.

Residual risk information is what is needed for the Health and Safety File – not risk assessments!

8.2 Residual Risks in Notifiable Projects

On Notifiable Projects, residual risk and other relevant information has to be provided by Designers, Clients and Contractors and given to the CDM-C for inclusion in the Health and Safety File for the structure. As the CDM-C is required to prepare, review and update the Health and Safety File, the responsibilities of Designers will generally relate only to providing sufficient information to the CDM-C in accordance with the agreed project programme. The CDM-C will then arrange for compilation of the information and the preparation and delivery of the Health and Safety File to the Client at the end of the construction phase of the project.

8.3 Residual Risks in Non-Notifiable Projects: Implications for Designers and Clients

On Non-Notifiable Projects, however, the situation is more complex for Designers and their Clients. Under the Regulations, Designers have to provide information to Clients to enable them to comply with their duties. Because there is a requirement for Clients to keep Health and Safety Files up to date then Designers will need to provide Clients with the relevant File related information from Non-Notifiable Projects that occur subsequent to a Health and Safety File being developed under the new CDM regime. This is an issue that Designers need to take on board, include in their procedures for all projects, and make sure that any other Designer/sub-consultant Designers they engage or interface with are aware of this requirement.

As noted earlier, Contractors are under no such obligation but Clients nevertheless need any information about residual risks that could be used to update a Health and Safety File and so where a Health and Safety File exists, information will have to be provided by Contractors to enable the Client to do this. Designers will need to advise Clients about this and also to advise them to take steps to make provision of this information a contractual obligation of all Contractors on such projects.

Whether there is a Health and Safety File or not, Designers and Contractors should provide their Clients with any information that the Client may need to provide to those planning and managing future works on their structures. Not to do so will prevent the Client from meeting his obligation under the CDM and other Regulations. It will, in the view of the APS, in due course become normal practice for Clients to build into conditions of engagement a requirement for Contractors (and Designers) to provide such information.

In the meantime, Designers can assist Clients by adopting procedures that ensure that this information is provided on all Non-Notifiable Projects. There may be little information, or none – but the Client will need what there is to be able to keep the Health and Safety File up to date. In due course this will benefit later Designers and Contractors, as well as those who have to work on the project, structure or site.

8.4 Client Responsibilities and Health and Safety File Information

Under the CDM Regulations, the Client has to keep the Health and Safety File available for inspection by anyone who may need it to comply with statutory provisions when working on the related structure or site; to revise it as often as may be appropriate to incorporate any new information; and to hand it over to anyone who acquires the structure or site, ensuring as he does so that they understand the nature and purpose of the File.

These generate three sets of consequences for Designers:

Client responsibilities: These are determined, not just by the CDM Regulations and the Health and Safety at Work Act 1974, but also by the Management of Health and Safety at Work Regulations, 1999. These state, with regard to risk assessments, that:

3. (1)   Every employer shall make a suitable and sufficient assessment of:

a) The risks to the Health and Safety of his employees to which they are exposed whilst they are at work and

b) The risks to the Health and Safety of persons not in his employment arising out of or in connection with the conduct by him of his undertaking...

This therefore requires that a Client has sufficient information always to be able to carry out that risk assessment (or have someone carry it out for him) – regardless of the type of project (Notifiable or Non-Notifiable). The implication is that the Client should always have a file of information that may be relevant to such assessments and that, where a project is Non-Notifiable, the Client should require those working on it to provide any and all information that is relevant to such assessments. Some simple examples here would be the discovery of asbestos (not then removed) that needs to be entered into the Client's written Asbestos Management Plan or Asbestos Register, or the use of a material that may be harmful if drilled or sanded, but there are many other risk issues about which a Client might need to receive information. This requirement applies to existing Files created under CDM 1994 and new Health and Safety Files established under the new CDM Regulations, as well as when there is no Health and Safety File at all.

Designers will therefore need to:

- *Ensure that residual risk and other Health and Safety information passed to the CDM-C (on Notifiable Projects) for the Health and Safety File is sufficient to enable appropriate risk assessments to be carried out at later date*
- *Advise Clients that they need the same sort of information to be provided by Designers and Contractors when working on Non-Notifiable Projects*

This suggests that, while not a requirement under CDM Regulations, there may be a need on all Non-Notifiable Projects to prepare a 'nascent' Health and Safety File or its equivalent so that any information relevant to the Health and Safety of those subsequently working on the structure or site can be readily located and taken into account by Clients and those working for them. This 'nascent' Health and Safety File will also be of use should any subsequent Notifiable Project be initiated – as a starting point for provision of pre-construction information.

Revision or Updating of the Health and Safety File: The requirement for a Client to revise the Health and Safety File after a project has been completed is simple where subsequent projects are Notifiable – development of the existing File will be part of the ongoing process. However, where subsequent works are Non-Notifiable, relevant information will be needed from Designers and Contractors and the Client will need to be advised to make that a condition of engagement – *this advice will almost certainly fall to the Designer to provide.*

Another issue that will need to be addressed by the Client is how the information is to be included in an existing File. Firstly it will need, for all but the simplest Files, to be located in the right place, cross-referenced as appropriate and integrated into the format of the File (for instance, by being delivered as appropriate electronic data where the Health and Safety File is in an electronic database system). Secondly, any interactions with other existing hazards and risks will need to be considered so that unexpected consequences of the new on the old (or vice versa) are understood. This suggests that the Client needs to decide who should deal with this updating or revision of the Health and Safety File so that it is dealt with competently.

Designers will therefore need to advise the Client not only on the need to obtain this information if it arises in a project, but also to advise that a competent person should carry out that revision. This might be the Client, the Designer, or someone else specifically appointed by the Client to deal with it.

In this way, the Clients will receive from all construction projects the information that will enable them to comply with their duties under the CDM and the other relevant Health and Safety regulations.

Where Non-Notifiable Projects produce little or no residual risk or other Health and Safety or operational information, there will be no need to produce anything other than simple information. It can be extremely useful to identify clearly that a structure has no significant risks associated with it. This will prevent later Designers or Contractors searching for information that does not exist. On larger complex Non-Notifiable Projects (for example, large-scale partial or total demolitions) there may be a need for a more structured approach. Circumstances will dictate this, not the application of a bureaucratic procedure; responses and actions should be appropriate and proportionate.

Handing the Health and Safety File to another: Where a Health and Safety File is prepared on a Notifiable Project, the CDM-C will explain the nature and purpose of the File to the Client and, when the File is handed over to another, the Client will be in a position to explain what it is for, how it is structured and how it can be used. This is best achieved by a brief written explanation of the nature and purpose of the File within the File that is handed over. Should the Client's staff change, or memories fade, then the written word will remain.

However, when residual risk or Health and Safety information is given to the Client on a Non-Notifiable Project, whether or not in the form of a nascent Health and Safety File, the Client will need to know what that information is, what it relates to and its relevance to others who may later acquire the structure or premises. Whilst there may be no duty for the Client to pass on this information (unless it has been added to an existing Health and Safety File established as a result of a Notifiable Project), it would clearly be good practice to do so.

When handing over relevant information on Non-Notifiable Projects, Designers might need to explain all of this so that the information can be used to inform anyone who subsequently works on the structure or site, or both. This can include, of course, those working on later Non-Notifiable Projects – a whole series of which could generate a significant amount of relevant Health and Safety or risk related information.

8.5 Critical Designer Issues on Non-Notifiable Projects

From these discussions on Non-Notifiable Projects it can be seen that Designers should be involved in:

- Advising Clients on the need for Designers and Contractors to provide them with residual risk and other Health and Safety-related information
- Making sure that Clients understand what this information is and its relevance when it is given to the Client

- Advising Clients to consider using a structured method of keeping that information available either on the current project or for a series of projects

- Advising Clients that, if they already have a Health and Safety File, they need to keep this up to date with information from Non-Notifiable as well as Notifiable Projects
- Making clear that information added to an existing File needs to be integrated into it and interactions between old and new risk information taken into account
- Advising that this should be carried out by someone competent to do that work

Simple Non-Notifiable Projects with few residual risks and little resultant Health and Safety information will require little from the Designer other than some basic advice to their Clients about such information and its importance and, later, some paperwork to pass on such information as there is. However, when more complex intense projects are envisaged – projects that could involve up to 16 or 17 people for 30 days – there could be much more to do. Clients may need assistance with the discharge of their responsibilities and assistance with keeping and organising the information.

Where significant Health and Safety information is being garnered from a Non-Notifiable Project, it might be sensible to suggest that this should be organised as a simple Health and Safety File so that, as projects take place, there is a consistent, useful and accessible source of information available to those who may need it. If and when a Notifiable Project occurs, this 'nascent' File would be the starting point for the provision of information by the Client.

Designers may well find themselves at the limit of what they want, or are prepared to do, on such projects, other than provide the advice that has been outlined above and pass on the information the Client needs. Advice to Clients might therefore be to appoint a person who normally works on Notifiable Projects as a CDM-C to provide the competent service that is needed for this work on the Non-Notifiable Project. This, of course, would tie in with earlier recommendations to consider such an appointment to assist the Client with making and maintaining management arrangements for the project and for ensuring that information is provided appropriately.

8.6   Information for Health and Safety Files

A Designer, or any other contributor of information to a Health and Safety File, can focus on the key issues and reduce paperwork and bureaucracy by keeping matters simple, and structuring the File for each structure to address just 11 questions or topics where this may be relevant to the Health and Safety of any future construction work. The level of detail should allow the likely risks to be identified and addressed:

(a) A brief description of the work carried out. Describe in more detail the steps taken to overcome any unanticipated significant problems and include information on methods of construction where special techniques were necessary, e.g. ground freezing, propping, dewatering

(b) Relevant information about access around and to the site, adjacent properties, businesses, processes, previous use, etc

(c) Relevant information on the ground conditions and other residual hazards and, where appropriate, how these have been dealt with (for example, surveys or other information concerning asbestos, contaminated land, water bearing strata, buried services)

(d) Key structural principles (e.g. bracing, sources of substantial stored energy, including pre- or post-tensioned members)

(e) Safe working loads for structural elements, e.g. floors, roofs, pile caps left in place, particularly where these may preclude placing scaffolding or heavy machinery there

(f) Hazardous materials used (e.g. lead paint, pesticides, special coatings which should not be burnt or ground off)

(g) Information regarding the removal or dismantling of the whole structure, or elements of the structure, or precast elements, as well as installed plant and equipment, bearings, joints, or other items with shorter life spans than the rest of the structure (e.g. lifting or jacking arrangements, weights, lifting points left in, centre of gravity of lifted elements, etc)

(h) Health and Safety information about equipment provided for cleaning or maintaining the structure

(i) The nature, location and markings of significant services, including fire-fighting services

(j) Information and 'As Built' drawings of the structure, its plant and equipment. Ensure that this information and/or drawings includes the means of safe access to and from service voids, fire doors and compartmentation

(k) Special features or precautions likely to be required if the structure is to be extensively modified or demolished (e.g. sequence of demolition or temporary works and anticipated loadings to avoid sudden or progressive collapse)

On projects where many individual structures are built, this type of file 'data sheet' format can greatly assist future maintenance and works planning for any individual structure.

'As Built' drawings: Construction drawings are often marked up with a few relevant notes from site, and the drawing revision changed to 'As Built'. This frequently leaves notes on the drawings that were entirely appropriate and necessary at the construction stage, but are worthless or even misleading for future works. Drawings must be reviewed, out-of-date or redundant and unnecessary notes and information removed, and information relevant to future works included. Where the 'for construction' drawing showed a necessary sequence of works, this must be translated into a safe sequence of works for removal or demolition of the element or structure. Where a Designer's assumed sequence of works was on a drawing, any changes or alternative means used by the Contractor must be translated into information to inform future Designers or Planners of works.

Use of the above 'data sheet' format, addressing the 11 issues, will reduce paperwork, and information provided in other hand-over documentation can be cross-referenced from this CDM-specific Health and Safety structure data sheet. It is also easy and helpful for the writer of the data sheet to include the words 'not applicable' or 'no significant issues/hazards' against any of the sections – and this will again warn the reader not to waste resources in a fruitless search for any other structure or project documentation.

8.7   Proportionality

The procedures and systems used on Non-Notifiable Projects to deal with residual information flow, organisation and records, should be closely related to the scale and nature of the issues involved and the amount of information that is likely to be dealt with so that the Client and Designer responses are proportionate to the issues involved. The last thing that is needed is unnecessary paperwork and bureaucracy, but equally there is a need to ensure that Non-Notifiable Projects – and the Health and Safety risks and issues inherent in them – are properly dealt with. The key is proportionality, and Designers will need to find the balance appropriate to each project if they take on this aspect of the work, or help the Client to do so if it is thought that another way of dealing with the issues would be more appropriate.

Records and Feedback

9.1 Why Keep Records?

The ACoP for the CDM Regulations makes it clear that Designers are not legally required to keep records of the process that they have used to try to achieve safe designs. However, accurate records of discussions are an important part of any project for many reasons. In matters related to Health and Safety, in particular, they demonstrate the thought processes used in reaching decisions and the issues that were relevant at the time that the decisions were made. This can be very important when designs are subject to change over, or after, a long period of time and where, perhaps, design co-ordination and change-tracking do not accurately identify such issues. There are also a number of occasions when a project or design is shelved or delayed for a significant length of time, and it may not necessarily be the same team of people who pick up the shelved or delayed project to again take it forward. Accurate records of the DRM process and decisions will greatly assist in this situation.

Properly kept records will also allow an audit trail to be developed, and this can be very useful in demonstrating that a decision taken at a certain time was a reasonable one at that time, which is what might have to be shown if misfortune strikes. It is also very helpful to a Designer under the stress of an investigation to have a record that can be used to refresh the memory and perhaps amplify diary notes and recollections.

It is also true to say that whoever has the better records has the better chance of disproving any allegations about shortcomings under CDM, and they might also help to demonstrate that attempts to discharge duties under CDM were frustrated by others or other considerations.

Confidentiality: In a Health and Safety case, the only information that does not have to be disclosed is discussions and correspondence between the Designer and legal counsel. Courts can impel Designers to make everything else available, even personal diaries and electronic exchanges. This suggests that care should be taken in all records, including personal diaries, to avoid 'jokey' or 'off-the-cuff' comments that could, at a later date, undermine the reliability, truthfulness or professionalism of a Designer's position.

9.2 Audit Trails and Records

Although there is no legal requirement for Designers to keep records, there are a number of reasons why they should do so:

- To create an auditable trail that can help to demonstrate that reasonable professional judgement has been exercised
- To record the process by which design decisions were eventually reached. This can help others at a later date to evaluate the possible impact of design changes
- As an aide memoire to help a defence in case of legal investigation
- To assist the design process on future similar projects

If in the future your design decisions are questioned by the HSE or Police, you may have difficulty recalling even the simplest of facts and will be grateful for well kept records

The records and audit trail do not have to be extensive or exhaustive, and the following would qualify:

- Letters sent and received
- E-mails
- Notes of meetings (internal and external)
- Quality management system records
- Health and Safety design management documentation
- File notes: these could be informal notes to remind you what influenced a particular decision
- Superseded drawings: only choose those with particular relevance

These would, of course, only be useful if they are properly referenced, dated and available.

It is always difficult to decide which information to keep, but it is worth bearing in mind the following:

- The information referred to in this Section applies to key decisions that might affect Health and Safety. Records will also need to be kept for other purposes
- Keep notes short – it should not always be necessary to record details of a discussion, only that it took place and the outcome
- Try to anticipate what information should be available in the event of investigations regarding the suitability and compatibility of designs (in terms of Health and Safety)
- If attempts to discharge duties were being frustrated by other people or influences, it might be useful for records to demonstrate this fact

Letters, e-mails, notes, etc: It should be obvious, but it is important to reiterate that Designers should not:

- Use language that is inflammatory or that deprecates the abilities of others
- Write personal comments on papers to be kept
- Record anything they would not wish others to see

and should:

- Stick to the facts
- Avoid unnecessary adjectives
- Remember that they may be required to produce any records kept, and they will be shown to others

Minutes of meetings: When minutes of meetings are received they must be checked for accuracy, both in terms of what was said, who was there and what actions were agreed. It is worth remembering that minutes tend to sum up discussions and, if they go unchallenged, they will become, by default, an accurate record of what was said. There will be no point in contesting minutes when misfortune strikes.

9.3   Keeping Records

What to keep: As well as the information that has been outlined above, some thought should be given to retention of full copies of Health and Safety information that has been handed over at critical times in a project, so that the hand-over position can be 'fixed' at some time in the future. This is particularly relevant to, for instance, information for the Health and Safety File that will be compiled by someone else (the CDM-C) and used by yet another person (the Client) who may immediately begin revision of it.

Storage: Information can be kept in any format as long as it can be extracted when required and cannot be altered. Paper copies are obviously bulky, so it may be worth considering having all information scanned and stored electronically. Records should be referenced so that specific information can be easily retrieved.

It is also worth remembering that, if information is stored electronically, it will be necessary to ensure that, when systems are updated, a means of recovering and reading that information in the future is retained. With the massive changes in electronic storage formats that have occurred in recent years, this may mean holding onto hardware as well as software.

How long for? There is no clear advice available regarding the length of time that information should be retained. Ideally, vital information should be kept until after a building is demolished; however, space is always a problem so, unless information can be kept in a purely electronic format, organisations will need to make judgements about which information to retain and for how long.

9.4   Feedback

Learning from experience, or feedback, is an important part of any process. It will help avoid making the same mistake more than once, and help to develop competence and capability for future projects. To make the same mistake twice could be classed as being unprofessional and could detract from an ability to demonstrate that reasonable professional judgement has been exercised.

Feedback also links with Section 3 of this Guide which is concerned with the ability to demonstrate capability and competence. Drawing into this record the experiences and feedback from projects can help to show that Designers are taking the development of DRM capability seriously.

Index